HYBRID SIMULATION: THEORY, IMPLEMENTATION AND APPLICATIONS

BALKEMA – Proceedings and Monographs
in Engineering, Water and Earth Sciences

Hybrid Simulation: Theory, Implementation and Applications

Editors

Victor Saouma
Department of Civil, Environmental and Architectural Engineering,
University of Colorado, Boulder,
Principal Investigator and Director
Fast Hybrid Testing Laboratory
George E Brown, Jr. Network of Earthquake Engineering Simulation

Mettupulayam Sivaselvan
Department of Civil, Environmental and Architectural Engineering
University of Colorado, Boulder
CO-Principal Investigator
Fast Hybrid Testing Laboratory
George E Brown, Jr. Network of Earthquake Engineering Simulation

CRC Press
Taylor & Francis Group
Boca Raton London New York

CRC Press is an imprint of the
Taylor & Francis Group, an **informa** business
A TAYLOR & FRANCIS BOOK

CRC Press
Taylor & Francis Group
6000 Broken Sound Parkway NW, Suite 300
Boca Raton, FL 33487-2742

First issued in paperback 2019

© 2008 by Taylor and Francis Group, LLC
CRC Press is an imprint of Taylor & Francis Group, an Informa business

Typeset by Vikatan Publishing Solutions (P) Ltd., Chennai, India

No claim to original U.S. Government works

ISBN-13: 978-0-415-46568-7 (hbk)
ISBN-13: 978-0-367-38739-6 (pbk)

Library of Congress Cataloging-in-Publication Data

Hybrid simulation : theory, implementation and applications / edited by V. Saouma,
M.V. Sivaselvan.
    p. cm.
  Includes bibliographical references and index.
  ISBN 978-0-415-46568-7 (hardback : alk. paper) – ISBN 978-0-203-89294-7 (ebook)
1. Structural engineering–Computer simulation. 2. Hybrid computer simulation.
I. Saouma, Victor E. II. Sivaselvan, Mettupalayam V.

TA641.H93 2008
624.101'13–dc   2008010199

**Visit the Taylor & Francis Web site at
http://www.taylorandfrancis.com**

**and the CRC Press Web site at
http://www.crcpress.com**

Table of Contents

Foreword

As Shakespeare's Juliet famously mused to Romeo, *"What's in a name? That which we call a rose, by any other name would smell as sweet."* The title of this volume, on the other hand, will surely evoke alternative connotations in readers with different technical backgrounds, for it is not the *name* which differs across fields but the technologies behind that name.

A cursory search on the Internet reveals that techniques referred to as *hybrid simulation* have been in use for quite some time in diverse areas of knowledge, for example, in computer science, computer graphics and animation, robotics, control theory, and in bioinformatics. A common thread in all of these uses seems to lie in the combination, coordination and synchronization of discrete event simulations in a digital computer with external processes that loop back via continuous, analog signals. Similarly, hybrid simulation in this monograph refers to slow —and not so slow— seismic tests conducted on *real* specimens in coordination with *virtual* structural models running *in silico*. The latter are used to divine the inertia (and other) forces that the structure would have been exposed to during the actual test earthquake and affect in turn the progression of the experiment. Still, in earthquake engineering, the now widespread use of the appellative *hybrid simulation* in lieu of the previously common *pseudodynamic testing* seems to be a relatively new development which took hold mainly in the new millennium. Indeed, a comprehensive review of the numerous references in this work reveals that the word "hybrid" appears in titles of articles dated before 2000 only twice.

My own contribution to hybrid simulation came in the form of two companion articles in the May 1998 issue of the *Journal of Engineering Mechanics*, where I proposed an as yet untried hybrid testing method to endow shaking tables with thrust devices in the form of actuators, unbalanced flywheels or jet engines that act directly onto parts of the test specimen. In principle, this strategy would allow decomposing the seismic excitation into two arbitrary components, for example into low and high frequencies, with one assigned to the shaking table and the other to the actuators. The main benefit is a reduction in the stroke needed for the actuators, the speed at which these must react, or the maximum displacement that the table must accommodate. Alternatively, one could also supplement existing tables capable of moving in only one direction with thrusters that would allow simulating fully three-dimensional earthquake forces. At this point in time, however, the promise of that proposal remains dormant and awaits its actual verification in the laboratory.

This monograph covers a wide spectrum of topics that range from detailed descriptions of testing facilities, modern hardware, methods and algorithms for control, real time multi-directional testing, distributed and networked testing, to the application of hybrid simulation in the Civil, aerospace and motor vehicle industry. It documents the evolution of the state of the art emanating from the coordinated research efforts under the aegis of NEES in the US, which has supported in turn the development and placement in service of ever larger and more powerful hardware. These investigations will not only help us gain a better understanding of the response of structures to dynamic loads, but will prove essential in our search for new materials and in the development of devices for the dissipation of seismic (or wind) energy. As more knowledge is gained through testing, we shall also be in a better position to narrow the gap of understanding between the results of experiments under controlled, ideal laboratory conditions, and actual observations on the response of engineered structures to earthquakes. This is, of course, because the mechanical and material properties of real structures are seldom known with any degree of precision, and it is often difficult to extrapolate results from the laboratory in ways that account properly for the large dimensions and degree of complexity of actual structures, not to mention inelastic behavior elicited by large seismic deformations. Nonetheless, after browsing through this volume one is left with the sense that experiments are now possible whose scale and size would have been only a dream yesterday. And if the past is any indication, the best is yet to come.

Eduardo Kausel
Massachusetts Institute of Technology
Cambridge, Massachusetts

Introduction

Hybrid simulation, also known as hardware-in-the-loop simulation, or virtual prototyping, is a technique of combining physical and virtual components to study the dynamic behavior of complex engineering systems. Models are created that consist of two parts actively interacting during the simulation, (a) a physical subsystem—an experimental component representing a portion of a system and (b) a virtual subsystem—a computer model of the remainder of the system. One is reminded of the thought experiments used in elementary thermodynamics to introduce heat engines—a system that undergoes the cycle is coupled in succession with different heat and work reservoirs. There is thus some intrinsic appeal in being able to couple a system with different boundaries to create interesting behavior. This appeal is evidenced in the wide range of applications described in this monograph, ranging from civil engineering to aerospace and automobile engineering.

The original conception of hybrid simulation in civil engineering was that of replacing one or more elements in a finite element model by physical elements. The intention then was that the physically constructed elements would have no rate-dependent behavior (viscous, inertial) and that all such effects would be represented in the computer model. This form of simulation was therefore termed *substructure pseudo-dynamic testing*. Since the physical components did not have any rate-dependent effects, the simulation could be performed arbitrarily slowly. Actuator dynamics was therefore not an issue, and the only extraneous features of concern were measurement noise and disturbance. These were considered in the same light as numerical errors due to round-off and truncation. Much progress was then made on developing numerical integration schemes for hybrid simulation with emphasis on error propagation analysis.

More recently, there has been growing interest in *real-time substructuring*, with the goal of representing rate-dependent effects in the physical subsystems, in particular inertia effects in some cases, accurately. This becomes especially important in aerospace and automobile applications where the distributed inertia of components plays a significant role. There have also been applications involving dissipation devices in seismic engineering, where real-time hybrid simulation has been found to be useful. In this form of hybrid simulation, actuator dynamics need to be explicitly considered. A number of *delayed-compensation* schemes have been developed for this purpose. Even more recently, the feedback structure of real-time hybrid simulation has been recognized, and a number of concepts from control systems theory are being applied. In particular, this approach provides a framework for the analysis of stability and robustness of hybrid simulation.

The purpose of this monograph is to provide a flavor of the different theoretical developments, implementations and applications of this interesting technique. It should be of particular interest to Civil, Aerospace and Automotive Engineers interested in exploiting a truly innovative testing paradigm which has not yet seen its full capabilities exploited.

Mettupulayam Sivaselvan
Victor Saouma
Boulder, CO
January, 2008

Theory

CHAPTER 1

Hybrid simulation: A historical perspective

M. Nakashima, J. McCormick & T. Wang
Disaster Prevention Research Institute, Kyoto University, Kyoto, Japan

ABSTRACT: The concept of hybrid simulation in structural engineering dates back to the late 1960s and early 1970s. Over the following decades, several developments and expansions of hybrid simulation were explored. These developments benefited from advances in control systems, electronics, computing, and mechanics among other areas. Experimental error propagation and the use of different integration algorithms were some of the first challenges that were addressed. One of the most significant expansions was the combination of substructuring techniques with hybrid simulation which made it possible to consider distributed hybrid simulation and real-time hybrid simulation. Within structural engineering, the advantages and drawbacks of hybrid simulation need to be considered against other experimental methods such as quasi-static and shake table testing. However, the benefit of being able to combine both experimental and analytical techniques provides a versatile means for performance evaluation of structural systems.

1 INTRODUCTION

Studies involving the behavior of structural systems under seismic loadings typically fall within either the experimental or analytical domain. Over the years, significant strides have been made in both areas. These advances have been further enhanced by various innovations in the area of control systems, computing, electronics, and mechanics. Many consider shake table tests to be the most realistic means of evaluating the performance of a structure under actual seismic loads. However, hybrid simulation provides an efficient alternative for evaluating the performance of structural systems combining both analytical and experimental techniques. Since conception, hybrid simulation has also been known as the online test or the pseudo-dynamic test.

Hybrid simulation is a test method that reproduces seismic loads experienced by a structure through the use of experimental techniques employing either quasi-static jacks or hydraulic actuators combined with numerical simulation. As a result of using mathematical methods to account for the dynamics effects on the structure and to provide the loading history imposed on the test specimen, quasi-static loading systems can be used in the experimental portion of the simulation. The earliest work on the development of hybrid simulation began in Japan and soon after in the United States. Major advances have been made in terms of understanding experimental error, integration algorithms, substructuring techniques, and real-time loading over the years. Many of these advances are in part a result of advances in other fields.

Along with the evolution of hybrid testing and its current status within structural engineering, the advantages and drawbacks of this technique for the evaluation of structural behavior under seismic loads is important to consider. The advantages and drawbacks need to be understood within the context of the role which experimental testing plays within structural engineering and the need to evaluate not only the capacity of structures, but also their performance given the current focus on performance-based design within the structural engineering community.

2   EARLY EVOLUTION OF HYBRID SIMULATION

2.1   *Initial development of hybrid simulation*

The concept of hybrid simulation dates back to the late 1960s when it was first proposed by Japanese researchers (Hakuno et al., 1969). A single-degree-of-freedom system was analyzed under seismic loadings using an analog computer in order to solve the equations of motion and an electromagnetic actuator was used to load the structure. This conceptual idea of hybrid simulation provided promise of an alternative means to gain important information on the behavior of structures under seismic loadings without the use of a shake table. The first major step in the development of the hybrid simulation method came about with the introduction of the digital computer and the use of discrete systems. In the mid 1970s, Takanashi et al. (1975) established the hybrid simulation method in its present form. By studying the structural system as a discrete spring-mass system within the time domain, real-time loading of the structure became unnecessary eliminating some of the earlier control problems. This allowed hybrid simulation to work with typical quasi-static loading systems and provide the necessary basis to apply hybrid simulation to structural engineering. Figure 1 provides the basic procedure for implementing the hybrid simulation method.

During the late 1970s, 1980s, and early 1990s, efforts in the Japan and the United States were undertaken to expand upon and validate the hybrid simulation test method. These early efforts are outlined in Takanashi & Nakashima (1987), Mahin et al. (1989), and Shing et al. (1996). The adoption of hybrid simulation in other countries around the world was more gradual and did not occur in significant numbers until the mid- to late 1990s. The incorporation of digital-analog loading control, electro-servo controllers, and the microcomputer increased the efficiency of the hybrid simulation method. A number of early application tests using these revealed a variety of areas that needed to be addressed such as experimental error propagation and proper choice of integration algorithms to ensure stability during the hybrid simulation.

2.1.1   *Experimental error propagation*

Experimental error can be introduced into hybrid simulation as a result of experimental hardware and improper alignment of the test setup. This error can affect the hybrid simulation through a difference in the applied displacement versus the computed displacement or incorrect force measurements from the tested structure, which is then used to solve the equation of motion. These errors can be further complicated during digital-analog conversion depending on the hardware being used. Since hybrid simulation is a closed-loop system and a stepwise process, these errors can accumulate resulting in an overall decrease in the accuracy of the hybrid simulation. Shing & Mahin

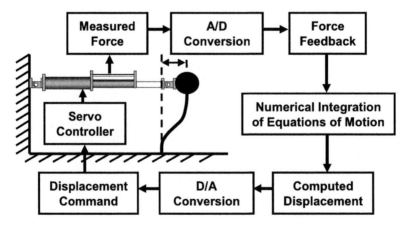

Figure 1.   Schematic of basic procedure for implementing hybrid simulation.

(1983), Kato et al. (1985), Nakashima et al. (1985), and Thewalt & Mahin (1987) provided significant contributions to identifying and determining the characteristics of experimental errors within hybrid simulation tests.

It was found that systematic overshoot error increases the apparent damping of the system. Systematic undershoot results in negative damping and can produce an increase in the response, particularly corresponding to the higher frequency response modes (Shing and Mahin, 1983, Kato et al., 1985, Shing and Mahin, 1987). Kato et al. (1985) found that displacement error is always undershoot error for quasi-static loading systems, but the response error does decrease with increased inelasticity. In order to minimize the effect of undershoot error, algorithms to suppress their effect or more precise displacement control can be used. In general, the displacement control needed for a hybrid simulation test is much greater than that required for conventional quasi-static tests given that the force values are used to calculate the next displacement increment. Further, the results of these studies made it clear that for stiff systems, the accuracy of hybrid simulation was somewhat limited.

2.1.2 *Integration algorithms*

The initial concept proposed by Takanashi et al. (1975) used the linear acceleration method to solve the difference equations. However, this required that the tangent stiffness of the test specimen be measured in order to form the equations of motion. For systems undergoing nonlinear behavior and stiff systems, this approach can result in difficulties due to the accuracy of the displacement sensors and loading equipment. As a result, Tanaka (1975) adopted the central difference method as the integration scheme for hybrid simulation. Since the central difference method is an explicit integration method, the tangent stiffness is not required and the reaction forces obtained from the tested structure can be used directly in solving the equations of motion for the next displacement step. Since these early tests, a variety of integration schemes have been considered for hybrid simulation in order to increase efficiency when considering multi-degree of freedom systems and also to reduce experimental error propagation. Much of the early development in regards to the use of various integration schemes for hybrid simulation can be found in Shing et al. (1996).

Integration schemes tend to fall into two categories: explicit or implicit. Explicit integration schemes have been found to be conditionally stable and require that systematic error be controlled within given parameters in order to remain stable. A single-degree-of-freedom test on a steel shear panel conducted by Nakashima et al. (1995) showed the feasibility of using explicit integration. However, explicit integration schemes may not be as feasible for multi-degree-of-freedom structures due to systematic undershoot errors exciting higher mode responses. As a result, implicit integration schemes which are unconditionally stable and can better handle structural nonlinearity and experimental error have been considered. However, implicit integration schemes often require iterations which can cause problems with tested specimens. The Operator-Splitting method is one unconditionally stable integration scheme considered by Nakashima et al. (1990) for hybrid simulation. One of the benefits of the Operator-Splitting method is that it does not require iterations for nonlinear analysis and thus avoids problems with path dependent effects on the experimental specimen. Shing et al. (1991) and Shing & Vannan (1991) showed the ability to apply the α-method (Hilber et al., 1977) with a modified Newton method to multi-degree-of-freedom systems successfully. The early work done with respect to integration algorithms led the way for substructuring and fast hybrid simulation.

2.2 *Substructuring in hybrid simulation*

One major advance in hybrid simulation was the use of substructuring techniques. Before the mid-1980s, most applications of hybrid simulation required testing of the complete structural system being considered. Thus, these tests could be expensive and require a large-scale testing facility. By using substructuring techniques typically applied to conventional dynamic analysis, the complete structure can be separated into several pieces. As a result, the parts of a structure that experience complex behavior, which may be difficult to accurately model numerically, are tested physically, while those parts of the structure which have a consistent behavior and are well defined are analyzed numerically. Thereby, substructuring reduces the space requirement in order to perform hybrid simulation tests and increases the ability to look at specific local component behavior. A schematic showing the concept of substructuring is shown in Figure 2.

Although the advantages of substructuring in hybrid simulation were clear, initial difficulties existed in applying it to general structures. The first major problem concerned numerical integration, since substructuring increased the total number of degrees of freedom. The increase in the number of degrees of freedom required a reduction in the time step in order to ensure stability when using the central difference method. A small time interval resulted in smaller displacements which are more difficult to control. Thus, these small time intervals can increase the propagation of systematic errors. The second problem resulted from having to make imaginary cuts in the structure. For substructuring, cuts must be made so that the physical specimen can be feasibly tested with the correct boundary conditions. This may revive rotational degrees of freedom which are difficult to control. In order to address these problems, Nakashima & Takai (1985) carried out studies to determine feasible integration techniques for substructuring and determined the applicability of mixed explicit-implicit integration algorithms. Similar studies in the United States by Dermitzakis & Mahin (1985) and Thewalt & Mahin (1987) dealt with special considerations required for massless degrees of freedom and mixed integration algorithms. By the early 1990s, hybrid simulation tests using substructuring techniques had been successfully carried out (Nakashima et al., 1990 & Shing et al., 1994).

2.3   Real-time hybrid simulation

The hybrid simulation test does not require dynamic loading as dynamic effects are accounted for in the numerical solution of the equations of motion. However, the development of velocity dependent structural components and devices to control the response of structures has led to interest in expanding the capacity of hybrid simulation to work in real-time. Focus on real-time hybrid simulation began in the early 1990s and has continued to present as more velocity dependent systems are conceived and applied to structures. The major difference between real-time hybrid simulation and pseudodynamic hybrid simulation is that velocities as well as displacements are controlled for the experimental portion of the test. An overview of the development of real-time hybrid simulation systems is provided by Nakashima (2001). The earliest real-time hybrid simulation system was applied to a single-degree-of-freedom system using a single dynamic actuator (Nakashima et al., 1992). This system required the development of a digital servomechanism in order to ensure accurate displacement and velocity control. However, given constraints with electronics, only a single-degree-of-freedom system could be considered.

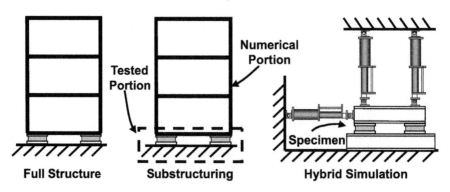

Figure 2.   Schematic of the concept of substructuring for a base isolated structure.

Developments in the electronics field paved the way for the development of the real-time hybrid simulation system in the late 1990s. Nakashima & Masaoka (1999) made further improvements to the earlier hybrid simulation system proposed by Nakashima et al. (1992) to take advantage of digital servo controllers and a newly devised computer to generate the target displacements and displacement signals for the experimental hardware. This computer was a key component to the real-time hybrid simulation in that it separated the 'response-analysis task' (solving of the equation of motion and obtaining the target displacement) from the 'signal-generation task' (creation of the successive displacement signals). This procedure was what allowed for continuous real-time loading without interruption of the displacements signals sent to the digital controller. Other early studies of real-time hybrid simulation suggested (Thewalt & Mahin, 1987) the possibility of a force controlled, real-time hybrid simulation.

These initial studies into the development of real-time hybrid simulation showed that there are some limitations. Since dynamic actuators are used, the size of the specimen is somewhat limited. Further, given the need to apply loads quickly and the effect that systematic error can have on the accuracy of the structural response, real-time testing may be severally limited in use with stiff structures. These initial tests also only dealt with single-degree-of-freedom systems. More difficulties are likely to arise as a result of controlling multiple actuators needed for multi-degree-of-freedom tested structures. As a result, many of the current hybrid simulation studies have focused on resolving some of these limitations.

3   CURRENT STATUS OF HYBRID SIMULATION

Hybrid simulation has come a long way since its inception. Its popularity amongst structural engineering researchers outside of both Japan and United States has also grown. Distributed testing is one recent concept that has developed from the use of substructuring and benefited from technological advances in data transfer and computing (Tsai, 2003). The concept of distributed testing is that individual substructures do not need to be within the same facility, but can be linked by either the internet or another means of data transfer. This concept benefits the experimental portion of hybrid simulation because a number of laboratories can be used providing a much larger capacity and the ability to take advantage of a larger variety of systems. There are also benefits in terms of the numerical portion of hybrid simulation in that the computers running the analysis do not need to be in the laboratory allowing for the use of stronger computers or even supercomputing facilities to run the hybrid simulation test. Other recent advances have been focused on the further development of real-time hybrid simulation.

One recent pseudodynamic hybrid simulation system developed in Japan is the peer-to-peer (P2P) Internet online hybrid test system (Pan et al., 2006). This system takes advantage of substructuring techniques and evaluates them in parallel either through numerical analysis or experimental testing. One of the unique aspects of this system is that the substructures are treated as independent and highly encapsulated systems. This is a result of the equation of motion being solved independently for each substructure rather than for the full structure as has typically been done in past hybrid simulation systems. As a result, only boundary displacement values and corresponding forces need to be exchanged between the substructures and the coordinator program, which ensures compatibility and equilibrium. No communication is required between individual substructures allowing data exchange through a simple input/output interface over the internet. This further allows commercial finite element codes to be used within the P2P system. The coordinator program uses an iterative predicting-correcting quasi-Newton procedure in order to determine the compatible boundary displacements and ensure equilibrium. By using either the initial elastic stiffness or secant stiffness of the tested substructures during the iteration process, only a single physical loading for each time step is necessary. Thereby, path-dependent loading effects are avoided. A proxy computer along with a socket mechanism allows for data exchange over the internet ensuring the ability to perform distributed tests. A schematic of the P2P system can be seen in Figure 3.

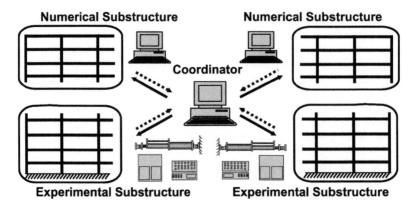

Figure 3.   Schematic of the peer-to-peer (P2P) internet online hybrid test system (Pan et al., 2006).

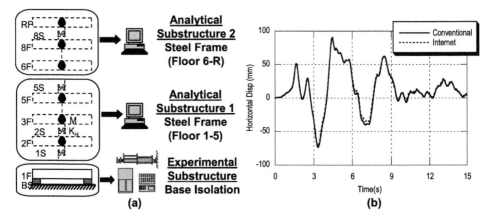

Figure 4.   (a) Substructuring for the base isolation verification study and (b) displacement time history comparison for the conventional hybrid simulation and the P2P online system (Pan et al., 2006).

A series of verification studies have been conducted to confirm the viability of the P2P Internet online hybrid test system. A base isolation system was used in order to compare the P2P system to a conventional hybrid simulation system in which the equations of motion were solved for the full structure (Pan et al., 2006). Figure 4a shows the substructuring for the hybrid simulation test. The maximum difference in the displacement time histories was only 5% suggesting good accuracy of the P2P system, Figure 4b. A second verification tests focused on the ability of the P2P system to handle more complex nonlinear behavior and multiple commercial finite element programs using a steel-encased reinforced concrete (SRC) structure shown in Figure 5a (Wang et al., 2007). Since only forces and displacements need to be passed between the substructures and coordinator in the P2P system, commercial finite element software can be employed without any modification to the source code by using the restart capability. The displacement time history for the first story of the steel tower is shown in Figure 5b for the P2P test (experimental substructure) and the case where the whole system is analyzed numerically. A final verification test was completed to ensure that the P2P system can work for multiple tested substructures in multiple locations (McCormick et al., 2007). A four-story steel moment frame was studied where the superstructure was taken as the analytical substructure and the plastic hinges at the column base were taken as the experimental substructures, Figure 6a. Figure 6b shows the complexity of the column-base behavior obtained during the P2P study and the advantage of using hybrid simulation to accurately obtain this behavior

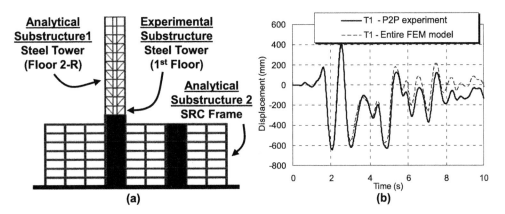

(a)

(b)

Figure 5. (a) Substructuring of SRC structure and (b) displacement time history (Wang et al., 2006).

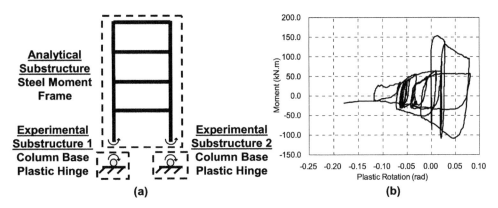

(a)

(b)

Figure 6. (a) Substructuring for the steel moment frame study and (b) moment-rotation relationship for one of the experimental column bases (McCormick et al., 2006).

from the experimental substructure. In general, the P2P Internet online hybrid test system expanded upon the advantages of conventional hybrid simulation methods. Other hybrid simulation systems have been developed in the United States including the Simcor system (Kwon et al., 2005) and OpenFresco (Takahashi et al., 2006).

Along with the development of various systems, current work has focused on improving real-time hybrid simulation. This work has focused on developing better integration schemes to reduce error propagation as a result of time delays associated with communication and loading or systematic error resulting from the loading equipment (Jung & Shing, 2006). However, there still exist many limits to the use of real-time hybrid simulation within structural engineering that continue to prevent its wide spread use.

Over the years, significant strides have been made in the development of pseudodynamic hybrid simulation, while there is still work needed in order to make real-time hybrid simulation applicable over a wide variety of structural systems. For pseudodynamic hybrid simulation, the largest need is further verification against full-scale shake table tests. Applications that truly take advantage of the benefits of hybrid simulation need also to be considered. For real-time hybrid simulation, many of the initial problems in its implementation are still being considered. Although real-time hybrid simulation provides an interesting challenge and opportunity to further the hybrid simulation method, it should not overshadow pseudodynamic hybrid simulation in future research efforts.

4   ROLE OF HYBRID SIMULATION IN STRUCTURAL ENGINEERING

The role of hybrid simulation within structural engineering is very much intertwined with the role of experimental testing. Experimental testing is necessary within structural engineering to confirm and develop the numerical models that are used in analytical software. Further, experimental testing can provide information of the actual behavior of structures in light of not having data of the performance of structures during an actual earthquake. This information is key to the development of design guidelines and new structural systems to resist earthquake loadings. In this regard, hybrid simulation provides the unique benefit of combining both analytical and experimental techniques particularly when substructuring techniques and distributed testing are employed. However, the question of the benefits and drawbacks of hybrid simulation versus other experimental testing techniques still remains. The benefits of pseudodynamic hybrid simulation and real-time hybrid simulation must be considered.

4.1   *Comparison of hybrid simulation versus conventional quasi-static loading tests*

Conventional quasi-static loading tests use a predetermined loading protocol in order to test the behavior of structural components and full-scale structural systems. In essence, the only difference between quasi-static loading tests and hybrid simulation tests is the loading protocol as the experimental hardware used for both is basically the same. Both types of experimental tests can provide a good idea of a structure's capacity and information in regards to the accuracy of numerical models as a benchmark study. However, the quasi-static loading technique is limited by the predetermined loading protocol, while hybrid simulation can be used to further look at the complex behavior of a structure which may result from an earthquake loading. As a result, hybrid simulation provides a better idea of the performance of a structure during an earthquake, which is important given the focus of the earthquake engineering community on performance-based design. In general, both of these experimental tools provide an effective means of evaluating structures with quasi-static loading tests being useful for capacity analysis and hybrid simulation providing information on performance assessment.

4.2   *Comparison of hybrid simulation versus shake table tests*

The most significant advantage that hybrid simulation has over shake table tests is the fact that structures can be analyzed under actual earthquake loads with the load being applied at a quasi-static rate. This lower loading rate allows larger specimens to be studied, since the capacity of quasi-static jacks is commonly larger than that of dynamic jacks. The ability to test larger specimens is important when consideration is given to scaling effects. It is well known that the size of specimens can play a critical role in a structure's behavior and it is often difficult to properly scale all portions of a specimen. This is particularly true for connection details and even material properties. Scaled specimens can provide a good understanding of global behavior, but local behavior may not be able to be simulated accurately because of these constraints. However, this local behavior may play a critical role in determining the performance of a structure given that initial damage usually occurs on a local level. Slow loading also allows for observation of the behavior of various structural components, which cannot be done with shake table tests due to the real-time loading rate. The ability to be able to monitor a test continuously is a considerable benefit in detecting the initiation of local behavior and determining how damage in a structure progresses over time during an earthquake. This provides an incredible aid in teaching and developing new systems to limit structural response during an earthquake. Conventional equipment can be used with hybrid simulation allowing tests to be performed in almost any moderately sized laboratory. The ability to be able to perform hybrid simulation in most laboratories as a result of the use of quasi-static loading is beneficial, particularly in conjunction with distributed testing and substructuring. In this way, hybrid simulation can take advantage of several laboratories to test a large number of substructures in the process of evaluating the behavior of a single system.

Shake table tests provide the advantage of being able to load specimens real-time to account for any loading rate effects that may be present in a structure. However, most conventional structural materials do not show any significant rate dependence. Although, more and more rate dependent structural devices, such as fluid viscous dampers, are being developed for structural control. Recent advances have been made in shake table testing with the construction of the large-scale shake table facility at E-Defense in Miki city, Japan (Nakashima, 2006). The E-defense shake table can accommodate specimens up to a weight of 12 MN (1,200 metric ton) and thus removes scaling effects. However, this is the only known shake table that can tests mid-size, full-scale structural specimens and the cost and time required to perform such a test are much greater than those associated with hybrid simulation. The E-defense shake table does provide the benefits of being able to test full-scale systems under actual real-time seismic loadings and is invaluable to fully understanding the performance of structures. The capacity of conventional shake tables tends to be much more limited requiring that scaled structural models be tested. As was mentioned previously, it is often difficult to scale a prototype specimen perfectly. In general, the benefits of hybrid simulation versus shake table testing come down to a question of loading-rate effects versus scaling effects. Often times with conventional materials, it can be contended that testing full-scale specimens is of much greater importance than testing them at real-time loading rates.

4.3   *Pseudodynamic hybrid simulation versus real-time hybrid simulation*

Real-time hybrid simulation has a similar benefit to pseudodynamic hybrid simulation when compared to shake table tests in terms of the ability to test larger specimens. Although dynamic actuators, which have lower load-applying capacity compared to quasi-static jacks of the same size, are used, real-time hybrid simulation systems do not require the full mass of the structure in the experimental portion as this is accounted for in the analytical model. Substructuring techniques further decrease the size of specimens that need to be fabricated since only those portions of a structure which exhibit complex behavior or have loading-rate dependence need to be tested physically. The fact that numerical modeling is used for part of the structural system also makes it much easier to maintain similitude if scaling is necessary to some extent. However, shake table testing has some advantages with respect to hybrid simulation in that error propagation due to limitations in the loading system and time for computing are not present. Shake table tests are also viable for any type of structure where the application of real-time testing can be difficult, particularly for cases where substructures have multiple degrees-or-freedom or are very stiff.

In general, the benefits and drawbacks of pseudodynamic hybrid simulation versus real-time hybrid simulation are similar to those for the comparison of hybrid simulation versus shake table testing. The choice of which type of hybrid simulation test to use still depends on a comparison of loading rate effects versus size effects. Given the current state of real-time hybrid simulation, however, it is clear that pseudodynamic hybrid simulation is currently the more versatile.

5   CONCLUSIONS

The concept of hybrid simulation for testing structural systems under earthquake-type loads has seen significant expansion and improvement since its conception in the late 1960s and early 1970s. Efforts to understand experimental errors that limited its inception and the application of various integration algorithms have further improved the performance of hybrid simulation systems. The use of substructuring techniques with hybrid simulation was one of the most significant expansions as it led the way for current studies focused on distributed hybrid simulation and real-time hybrid simulation. As a result of all these developments, hybrid simulation has been shown to have significant benefits within structural engineering, particularly in the area of performance evaluation of structures. Even compared to shake table tests, hybrid simulation techniques provide some advantage depending on whether scaling or loading rate effects are a larger factor for the particular structural system being study. However, a few key points in regards to hybrid simulation need to be considered as future expansion and development of this testing technique is undertaken. The benefits of slow loading should be considered, particularly in determining the direction of future studies using hybrid simulation. Work is still necessary in terms of calibrating hybrid simulation studies against actual full-scale structures undergoing earthquake loadings. This need should not be lost in the pursuit of fast loading systems. Second, care must be taken in implementing substructuring techniques and choosing analytical substructures. Analytical substructures can only be justified if the behavior of these substructures can be modeled accurately or the hybrid simulation as a whole may be compromised. Finally, feasible applications of hybrid simulation need to be considered during future endeavors to ensure continued support of the engineering community.

REFERENCES

Dermitzakis, S.N. & Mahin, S.A. 1985. Development of substructuring techniques for on-line computer controlled seismic performance testing. *UBC/EERC-85/04, Earthquake Engineering Research Institute*. University of California, Berkeley, California.

Hakuno, M., Shidawara, M. & Hara, T. 1969. Dynamic destructive test of a cantilever beam, controlled by an analog-computer. *Transactions of the Japan Society of Civil Engineers*. 171: 1–9. (in Japanese).

Hilber, H.M., Hughes, T.J.R. & Taylor, R.L. 1977. Improved numerical dissipation for time integration algorithms in structural dynamics. *Earthquake Engineering and Structural Dynamics*. 5: 283–292.

Jung, R.Y. & Shing, P.B. 2006. Performance evaluation of a real-time pseudodynamic test system. *Earthquake Engineering and Structural Dynamics*. 35: 789–810.

Kato, H., Nakashima, M. & Kaminosono, T. 1985. Simulation accuracy of computer-actuator online testing. *Proc. of the 7th Symposium on the Use of Computers in Building Structures, Architectural Institute of Japan*. Tokyo, Japan: 199–204. (in Japanese).

Kwon, O.S., et al. 2005. A framework for multi-site distributed simulation and application to complex structural systems. *Journal of Earthquake Engineering*. 9: 741–753.

Mahin, S., et al. 1989. Pseudodynamic test method—current status and future directions. *Journal of Structural Engineering*. 115(8): 2113–2128.

McCormick, J. et al. 2007. Progress and applications of a peer-to-peer (P2P) Internet online hybrid test system. *Proc. of the 9th Canadian Conference on Earthquake Engineering*. Ottawa, Ontario, Canada.

Nakashima, M. 2001. Development, potential, and limitations of real-time online (pseudo-dynamic) testing. *Philosophical Transactions of the Royal Society of London A*. 359: 1851–1867.

Nakashima, M. 2006. Test on collapse behavior of structural systems. In J.J. Perez Gavilan E. (ed.), *Earthquake Engineering: Challenges and Trends*, Universidad Nacional Autonoma de Mexico, Mexico.

Nakashima, M., Akazawa, T. & Igarashi, S. 1995. Pseudo-dynamic testing using conventional testing devices. *Earthquake Engineering and Structural Dynamics*. 24(10): 1409–1422.

Nakashima, M., Kato, H. & Kaminosono, T. 1985. Simulation of earthquake response by pseudo dynamic (PSD) testing technique (part 3 estimation of response errors caused by PSD test control errors. *Proc. of the Annual Meeting of the Architectural Institute of Japan*. Tokyo, Japan: 445–446. (in Japanese).

Nakashima, M., Kato, H. & Takaoka, E. 1992. Development of real-time pseudo dynamic testing. *Earthquake Engineering and Structural Dynamics*. 21: 79–92.

Nakashima, M. & Masaoka, N. 1999. Real-time on-line test for MDOF systems. *Earthquake Engineering and Structural Dynamics*. 28: 393–420.

Nakashima, M. & Takai, H. 1985. Computer-actuator online testing using substructure and mixed integration techniques. *Proc. of the 7th Symposium on the Use of Computers in Building Structures, Architectural Institute of Japan*. Tokyo, Japan: 205–210. (in Japanese).

Nakashima, M., et al. 1990. Integration techniques for substructure pseudodynamic test. *Proc. of the 4th U.S. National Conference on Earthquake Engineering*. Palm Springs, California: 515–524.

Pan, P., et al. 2006. Development of peer-to-peer (P2P) Internet online hybrid test systems. *Earthquake Engineering and Structural Dynamics*. 35: 267–291.

Shing, P.B., Bursi, O.S. & Vannan, M.T. 1994. Pseudodynamic tests of concentrically braced frame using substructuring techniques. *Journal of Constructional Steel Research*. 29: 121–148.

Shing, P.B. & Mahin, S. 1983. Experimental error propagation in pseudodynamic testing. *UBC/EERC-83/12, Earthquake Engineering Research Institute*. University of California, Berkeley, California.

Shing, P.B. & Mahin, S.A. 1987. Cumulative experimental errors in pseudodynamic tests. *Earthquake Engineering and Structural Dynamics*. 15(4): 409–424.

Shing, P.B., Nakashima, M. & Bursi, O.S. 1996. Application of pseudodynamic test method to structural research. *Earthquake Spectra*. 12(1): 29–56.

Shing, P.B. & Vannan, M.T. 1991. Implicit algorithm for pseudodynamic tests: convergence and energy dissipation. *Earthquake Engineering and Structural Dynamics*. 20: 809–819.

Shing, P.B., Vannan, M.T. & Carter, E.W. 1991. Implicit time integration for pseudodynamic tests. *Earthquake Engineering and Structural Dynamics*. 20: 551–576.

Takahashi, Y. & Fenves, G.L. 2006. Software framework for distributed experimental-computational simulation of structural systems. *Earthquake Engineering and Structural Dynamics*. 35: 267–291.

Takanashi, K. & Nakashima, M. 1987. Japanese activities on on-line testing. *Journal of Engineering Mechanics*. 113(7): 1014–1032.

Takanashi K., et al. 1975. Non-linear earthquake response analysis of structures by a computer actuator on-line system (part 1 details of the system). *Transactions of the Architectural Institute of Japan*. 229: 77–83. (in Japanese).

Tanaka, H. 1975. A computer-actuator on-line system for non-linear earthquake response analysis of structures. *Journal of the Institute of Industrial Science*. 27: 15–19. (in Japanese).

Thewalt, C.R. & Mahin, S.A. 1987. Hybrid solution techniques for generalized pseudodynamic testing. *UBC/EERC-87/09, Earthquake Engineering Research Institute*. University of California, Berkeley, California.

Tsai, K., et al. 2003. Network platform for structural experiment and analysis (I). *NCREE-03-021, National Center for Research on Earthquake Engineering*. Taiwan.

Wang, T. et al. 2007. Numerical characteristics of peer-to-peer (P2P) Internet online hybrid test system and its applications to seismic simulation of SRC Structure. *Earthquake Engineering and Structural Dynamics*. (Accepted for Publication).

CHAPTER 2

Trajectory exploration approach to hybrid simulation

M.V. Sivaselvan
Department of Civil, Environmental and Architectural Engineering,
University of Colorado, Boulder, USA

J. Hauser
Department of Electrical and Computer Engineering,
University of Colorado, Boulder, USA

ABSTRACT: Hybrid simulation is a technique of combining physical and virtual components to study the dynamic behavior of complex engineering structures. Models are created that consist of two parts actively interacting during the simulation, (a) a physical substructure—an experimental test piece representing a portion of a structure and (b) a virtual substructure—a computer model of the remainder of the structure. The interface conditions between the two substructures are imposed by a transfer system. Here, we take the point of view of system trajectory (or behavior) exploration. We attempt to answer the question of whether the hybrid system consisting of physical and virtual components and the transfer system can exhibit certain trajectories (or behaviors) of the full system. In other words, we seek trajectories of the transfer system that are compatible with desired trajectories of the full system. In this article, we restrict ourselves to periodic trajectories and to the transfer system being a hydraulic uniaxial shaking table. When exactly compatible transfer system trajectories exist, computing them requires solving a two-point boundary value problem. Otherwise, we pose the question as a nonlinear least squares problem. We however do not address the solution of this least squares problem in this article. Taking a trajectory exploration approach may lead to more revealing ways of laboratory experimentation in structural dynamics in general.

1   TRAJECTORY EXPLORATION APPROACH TO HYBRID SIMULATION

Hybrid simulation, also known in other fields of engineering as hardware-in-the-loop simulation, virtual prototyping etc., is a technique of combining physical and virtual components to study the dynamic behavior of complex engineering systems. Models are created that consist of two parts actively interacting during the simulation: (a) a physical subsystem—an experimental component representing a portion of a system and (b) a virtual subsystem—a computer model of the remainder of the system. In this article, we take the point of view of system trajectory (or behavior) exploration. We consider a system as being modeled by a nonlinear differential equation,

$$\dot{x} = f(x, w) \tag{1}$$

where x is the vector of states of the system, and w is the vector of external disturbances, such as earthquake ground motion. In order to examine its behavior, we dissect the system, i.e., we partition its states as

$$\dot{x}_1 = f_1(x_1, x_2, w)$$
$$\dot{x}_2 = f_2(x_1, x_2, w) \tag{2}$$

We choose to build a physical representation of one of these subsystems, say x_1, which we consider more critical, or which wish to understand more about, and keep the other subsystem virtual. To stick

15

them back together, we need some *glue*, to get the state x_1 into the x_2 subsystem, and the state x_2 into the x_1 subsystem. This glue is provided by sensors, actuators, controls, and by an understanding of the system objectives and of the global nonlinear dynamics. Here for simplicity, we assume measurability of all physical states. The mathematical model of the system after it is glued back together is

$$
\begin{aligned}
\dot{x}_1 &= \bar{f}_1(x_1, x_a, w, u) \\
\dot{x}_2 &= f_2(x_1, x_2, w) \\
\dot{x}_a &= f_a(x_1, x_a, u)
\end{aligned}
\tag{3}
$$

which now includes the dynamics of the sensors and the actuators. x_a is the state of the actuators and sensors and u is a control input to the system. The differential equation for x_1 is also modified due to the interaction of the physical subsystem with the actuators. We approach hybrid simulation as the problem of determining trajectories of the hybrid system (3) that are compatible with trajectories of the full system (2).

We set out with a desired trajectory of the real system, $(x_{1,des}, x_{2,des}, w_{des})$. This trajectory may have been obtained, for example, by simulation of the mathematical model of the full real system, or from observation of a real-life structure, or from searching for a certain failure mode, or by some other means. We then ask if this trajectory is attainable by the hybrid system. Directly imposing the desired trajectory on the system (3) results in a differential algebraic equation (DAE) in x_a and u.

$$
\begin{aligned}
\dot{x}_a &= f_a(x_{1,des}, x_a, u) \\
0 &= \bar{f}_1(x_{1,des}, x_a, w_{des}, u) - \dot{x}_{1,des}
\end{aligned}
\tag{4}
$$

For periodic trajectories, this results in a two-point boundary value problem.

It is not necessary however, that the system (3) have trajectories that are compatible with those of the system (2), i.e. that the DAE (4) have a solution, because the actuator dynamics imposes additional constraints. We therefore would need to adopt an alternate, broader viewpoint. If a compatible trajectory does not exist, we can pose the question in a least squares sense as

$$
\min \|x(\cdot) - x_{des}(\cdot)\|^2 + \|w(\cdot) - w_{des}(\cdot)\|^2 + \|u(\cdot)\|^2
\tag{5}
$$

subject to the system dynamics of equation (3)

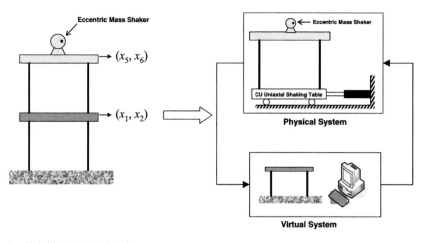

Figure 1.   Hybrid system explored.

This is a nonlinear least squares problem that can be solved, for example, using the *projection operator* approach (Hauser, 2002). Even if a minimum is not achievable, in searching for one, we explore the state space, and discover trajectories of the hybrid system that resemble the trajectories of the real system that we set out with. This alternate strategy is a subject of our current work.

In this article, we consider hybrid simulation with uniaxial shaking tables and look at periodic trajectories in particular. The system we consider, is a two-story building constrained to move in one plane. An eccentric mass shaker mounted on the top floor, provides periodic excitation to the building. We hybridize the system as shown schematically in Figure 1, keeping the top floor physical, and modeling the bottom floor virtually. We ask if this hybrid system, when excited by the eccentric mass shaker, can be made by means of suitable controls, to exhibit the same periodic trajectories as the full system.

2   MODEL OF A UNIAXIAL SHAKING TABLE

We first present a nonlinear model of the uniaxial shaking table. A schematic diagram of a shaking table driven by a hydraulic actuator fitted with a spool valve is shown in Figure 2(a). We follow the usual hydraulic models presented for example by Merritt (Merritt, 1967) and Jelali et. al. (Jelali and Kroll, 2003). However, in modeling flow through the servovalve ports 1–4, we pay particular attention to the orifice pressure-flow equation. Commonly, the equation obtained from Bernoulli's principle and applicable to high Reynolds number flows is used over the entire range of values of orifice pressure drop and valve opening. While this is a reasonable approximation since orifice flow is predominantly turbulent, it causes two mathematical difficulties in our trajectory exploration algorithms:

1. The resulting differential equations do not satisfy the Lipschitz condition when the pressure drop across the orifice is zero.
2. The pressure-flow equation is non-smooth when valve spool displacement is zero.

These difficulties however do not arise if we account for the following facts:

1. The orifice flow is *laminar* when the pressure drop or valve opening is small (i.e. low Reynolds numbers).
2. In *real* valves, there is clearance between the spool and the landings, so that there is some *leakage* flow even when the valve displacement is zero.

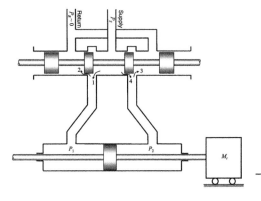

| (a) Schematic diagram | (b) Orifice area – spool displacement relationship |

Figure 2.   Hydraulic shaking table model.

With these considerations, we represent the orifice area as a function of the valve spool displacement with smooth transitions as shown in Figure 2(b). The smooth transitions may be obtained by fitting polynomials and matching any number of derivatives as desired with the straight line segments at either end of the interval. The orifice pressure-flow equation is given by

$$
Q(P, x_v) = \left(\sqrt{ C_d^2 A(x_v)^2 \frac{2}{\rho} P \tanh \alpha P + \left(\frac{\nu R_{tr}}{2b(x_v)} \right)^2 } - \frac{\nu R_{tr}}{2b(x_v)} \right) \tanh \alpha P
$$

$$
+ \frac{1}{2} \frac{C_d^2 A(x_v)^2 \frac{2}{\rho}}{\frac{\nu R_{tr}}{2b(x_v)}} (1 - \tanh^2 \alpha P) \tag{6}
$$

where P is the pressure drop across the orifice, x_v is the spool displacement, $A(x_v)$ is the area of the orifice, $b(x_v) = \frac{2w}{w^2 + A(x_v)}$, w is the orifice width, C_d is the turbulent flow discharge coefficient of the orifice, R_{tr} is the Reynolds number at which the flow transitions from laminar to turbulent, ρ is the density of the hydraulic oil, μ is the coefficient of viscosity of the hydraulic oil. α is a parameter such that the larger its value, the smaller the norm $||\mathrm{sgn}(x) - \tanh \alpha x||_1$. Equation (6) has been obtained by modifying an empirical flow equation proposed by Borutzky (Borutzky, Barnard and Thoma, 2002) such that $Q(\cdot, \cdot)$ is as differentiable as the smooth transitions in Figure 2(b).

The dynamics of the shaking table can be described by the following system of nonlinear differential equations.

$$
\dot{x}_1 = x_2
$$
$$
\dot{x}_2 = \frac{A_p}{M_t} x_3
$$
$$
\dot{x}_3 = f_3(x_1, x_2, x_3, x_4, u) \tag{7}
$$
$$
\dot{x}_4 = f_4(x_1, x_2, x_3, x_4, u)
$$

where

$$
f_3(x_1, x_2, x_3, x_4, u) = \frac{\kappa}{A_p(x_0^2 - x_1^2)} (x_0 Q_I(x_3, x_4, u) - x_1 Q_{II}(x_3, x_4, u) - 2K_l x_0 x_3 - 2A_p x_0 x_2)
$$

$$
f_4(x_1, x_2, x_3, x_4, u) = \frac{\kappa}{A_p(x_0^2 - x_1^2)} (x_0 Q_{II}(x_3, x_4, u) - x_1 Q_I(x_3, x_4, u) + 2K_l x_1 x_3 + 2A_p x_1 x_2)
$$

Here, the states x_1 to x_4 are respectively the displacement of the table, the velocity of the table the differential pressure between actuator chambers and the sum of the pressures in the actuator chambers. u is the control input, the spool displacement. It has been assumed here, that the spool displacement can be directly controlled, i.e., that the valve spool dynamics is much faster than the dynamics of interest. But our recent experiments have shown this not to be the case, and we are considering a more detailed model of the valve. Furthermore in equation (7), A_p is the piston area, M_t is the table mass, κ is the oil bulk modulus, x_0 is half the stroke of the actuator and K_l is the coefficient of leakage flow between the actuator chambers. Q_I and Q_{II} are combinations of flows in the four ports of the servovalve, and are given by

$$
Q_I(x_3, x_4, u) = Q(p_1, u) - Q(p_2, -u) + Q(p_3, u) - Q(p_4, -u)
$$
$$
Q_{II}(x_3, x_4, u) = Q(p_1, u) - Q(p_2, -u) - Q(p_3, u) + Q(p_4, -u)
$$

where p_i is the pressure drop across orifice i. Thus $p_1 = P_S - \frac{x_3+x_4}{2}$, $p_2 = \frac{x_3+x_4}{2} - P_R$, $p_3 = \frac{x_4-x_3}{2} - P_R$ and $p_4 = P_S - \frac{x_4-x_3}{2}$. $Q(\cdot, \cdot)$ is the orifice flow function (6).

3   FULL PHYSICAL SYSTEM MODEL

We model the full real system shown on the left side of Figure 1 by the following linear differential equations.

$$
\begin{aligned}
\dot{x}_1 &= x_2 \\
\dot{x}_2 &= -\frac{c_1+c_2}{m_1}x_2 + \frac{c_2}{m_1}x_6 - \frac{k_1+k_2}{m_1}x_1 + \frac{k_2}{m_1}x_5 \\
\dot{x}_5 &= x_6 \\
\dot{x}_6 &= \frac{c_2}{m_2}x_2 - \frac{c_2}{m_2}x_6 + \frac{k_2}{m_2}x_1 - \frac{k_2}{m_2}x_5 + \frac{1}{m_2}F(t)
\end{aligned}
\tag{8}
$$

where the states x_1 and x_2 are the displacement and velocity of the first floor and x_5 and x_6 are those of the second floor, m_i, c_i and k_i are the mass, damping and stiffness of floor i and $F(t)$ is the force from the eccentric mass shaker.

4   HYBRID SYSTEM MODEL

The model of the hybrid system shown on the top right of Figure 1 is

$$
\begin{aligned}
\dot{x}_1 &= x_2 \\
\dot{x}_2 &= \frac{A_p}{M_t}x_3 - \frac{c_2}{M_t}x_2 + \frac{c_2}{m_1}x_6 - \frac{k_1+k_2}{m_1}x_1 + \frac{k_2}{m_1}x_5 \\
\dot{x}_3 &= f_3(x_1, x_2, x_3, x_4, u) \\
\dot{x}_4 &= f_4(x_1, x_2, x_3, x_4, u) \\
\dot{x}_5 &= x_6 \\
\dot{x}_6 &= \frac{c_2}{m_2}x_2 - \frac{c_2}{m_2}x_6 + \frac{k_2}{m_2}x_1 - \frac{k_2}{m_2}x_5 + \frac{1}{m_2}F(t)
\end{aligned}
\tag{9}
$$

where the states, inputs and parameters are as described for the shaking table and for the full system.

5   TWO POINT BOUNDARY VALUE PROBLEM (TPBVP)
FOR PERIODIC TRAJECTORIES

Imposing the desired trajectory of the full system (8) on the hybrid system (9) leads to the condition

$$
x_{3,des} = \frac{M_t}{A_p}\left[(c_2x_{2,des} - c_2x_{6,des} + k_2x_{2,des} - k_2x_{5,des})\left(\frac{1}{M_t} - \frac{1}{m_1}\right) - \frac{c_1}{m_1}x_{2,des} - \frac{k_1}{m_1}x_{2,des}\right]
\tag{10}
$$

and the following DAE in x_4 and u:

$$
\begin{aligned}
\dot{x}_4 &= f_4(x_{1,des}, x_{2,des}, x_{3,des}, x_4, u) \\
f_3(x_{1,des}, x_{2,des}, x_{3,des}, x_4, u) &- \dot{x}_{3,des} = 0
\end{aligned}
\tag{11}
$$

Since the force $F(t)$ applied by the eccentric mass shaker is periodic, we seek periodic (steady state) solutions of this DAE. We thus solve the DAE with the boundary condition $x_4(T) = x_4(0)$. This leads to a TPBVP. We solve this by shooting with the ODE system

$$
\dot{x}_4 = f_4(x_{1,des}, x_{2,des}, x_{3,des}, x_4, u)
$$

$$
\dot{u} = -(D_5 f_3)^{-1} \left[\sum_{n=1}^{3} D_n f_3 \cdot \dot{x}_{n,des} - D_4 f_3 \cdot f_4 - \ddot{x}_{3,des} \right]
$$

$$
\dot{z}_4 = \left[D_4 f_4 - (D_5 f_3)^{-1} D_5 f_4 \cdot D_4 f_3 \right] \cdot z_4
$$

(12)

where we have used the shorthand f_i for $f_i(x_{1,des}, x_{2,des}, x_{3,des}, x_4, u)$ and D_i denotes the derivative of the function with respect to argument i. The third of these equations is the linearization of the first, which is needed for shooting using Newton's method. We solve the TPBVP with the following parameters of the MTS uniaxial shaking table and a two-story building model at the University of Colorado structures laboratory: $M_t = 2160$ lb, $A_p = 3.16$ in^2, $\kappa = 10^5$ psi, $x_0 = 9.75$ in, $P_S = 3000$ psi, $C_d = 0.6$, $\mu = 0.0067$ in^2/s, $\rho = 8.127 \times 10^{-5}$ lb-s^2/in^4, $R_{tr} = 10$, $w = 0.1656$ in, $m_1 = m_2 = 500$ lb, $k_1 = k_2 = 1464.5$ lb/in and c_1 and c_2 are such that there is 2% damping in the first mode of the full two-story structure. The desired trajectories of the full system and the compatible trajectories of the hybrid system are shown in Figure 3.

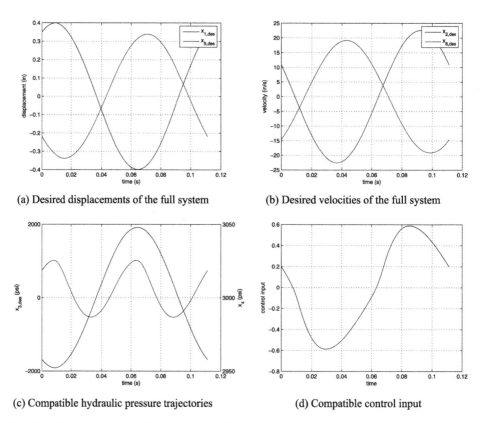

(a) Desired displacements of the full system

(b) Desired velocities of the full system

(c) Compatible hydraulic pressure trajectories

(d) Compatible control input

Figure 3.   Desired and compatible trajectories for the hybrid system.

6   SOLVING THE PERIODIC LEAST SQUARES OPTIMAL CONTROL PROBLEM

We wish to solve the following optimal control problem.

$$\min_{(x(\cdot),u(\cdot))} \int_0^T l(\tau,x(\tau),u(\tau))d\tau$$

$$\text{Subject to} \quad \dot{x} = f(x,u) \tag{13}$$

with the periodic condition $x(T) = x(0)$. l is periodic in its first argument. We consider two strategies.

6.1   *Strategy 1*

Following the idea outlined in (Hauser, 2003), we define the alternate optimization problem,

$$\min_{(x(\cdot),u(\cdot))} \int_0^T l(\tau,x(\tau),u(\tau))d\tau + \frac{1}{2}\rho^2 \|x(T) - x_0\|^2$$

$$\text{Subject to} \quad \dot{x} = f(x,u)$$

$$\text{with} \quad x(0) = x_0 \tag{14}$$

Let $(\bar{x}(\cdot), baru(\cdot))$ be the minimizer. $\hat{x} : x_0 \mapsto \bar{x}(T)$ defines a map. Let x_0^* be a fixed point of the map $\hat{x}$. Then solving the alternate optimization problem (14) with $x_0 = x_0^*$ gives the desired solution.

6.2   *Strategy 2*

Proceed from the following alternate definition of the projection operator. Let $\eta = (\alpha(\cdot), \mu(\cdot))$ be a periodic curve, i.e. $\alpha(0) = \alpha(T)$ and $\mu(0) = \mu(T)$. Define $\mathcal{P} : \eta \mapsto \xi$ where $\xi = (x(\cdot), u(\cdot))$ is given by

$$\dot{x} = f(x,u)$$

$$u(t) = \mu(t) + K(t)(\alpha(t) - x(t))$$

$$\text{with} \quad x(0) = x(T) \tag{15}$$

where $K(t)$ stabilizes periodic trajectories in the neighborhood of η. Note that projecting periodic curves results in a periodic control. Let $\xi = (x(\cdot), u(\cdot))$ be a periodic trajectory. Let $\mathcal{T}$ be the manifold of trajectories of the system $\dot{x} = f(x,u)$ with $x(0) = x(T)$. The following can be formally verified.

1. $\mathcal{P}(\mathcal{P}(\eta)) = \mathcal{P}(\eta)$ so that $\mathcal{P}$ is a projection.
2. Let $\zeta = (\beta(\cdot), \nu(\cdot))$ be a periodic curve. Then $D\mathcal{P}(\xi) : \zeta \mapsto \gamma$ where $\gamma = (z(\cdot), v(\cdot))$ is given by

$$\dot{z} = D_1 f(x,u) \cdot z + D_2 f(x,u) \cdot v$$

$$v(t) = \nu(t) + K(t)(\beta(t) - z(t))$$

$$\text{with} \quad z(0) = z(T) \tag{16}$$

3. Let $\gamma_1, \gamma_2 \in T_\xi \mathcal{T}$. Then $D^2 \mathcal{P}(\xi) : (\gamma_1, \gamma_2) \mapsto \omega$ where $\omega = (y(\cdot), w(\cdot))$ is given by

$$
\begin{aligned}
\dot{y} &= D_1 f(x, u) \cdot z + D_2 f(x, u) \cdot v + D^2 f(x, u) \cdot (\gamma_1(t), \gamma_2(t)) \\
w(t) &= -K(t) y(t) \qquad\qquad\qquad\qquad\qquad\qquad\qquad\qquad (17) \\
&\text{with} \quad y(0) = y(T)
\end{aligned}
$$

We continue to proceed formally. We consider the unconstrained minimization problem, $\min_\xi g(\xi)$ where g is defined as $g(\xi) = h(\mathcal{P}(\xi))$. We consider the computation of the Newton descent direction at $\xi \in \mathcal{T}$.

$$
\min_{\zeta \in T_\xi \mathcal{T}} Dg(\xi) \cdot \zeta + \frac{1}{2} D^2 g(\xi) \cdot (\zeta, \zeta) \tag{18}
$$

Note that $T_\xi \mathcal{T}$ is a linear vector space and not an affine space as is the case with fixed initial conditions. The only difference from the situation with fixed initial conditions which is caused by periodic conditions is in computing $Dh(\xi) \cdot D^2 \mathcal{P}(\xi) \cdot (\zeta, \zeta)$ for $\xi \in \mathcal{T}$ and $\zeta \in T_\xi \mathcal{T}$, which we now illustrate. Let $a(\tau) = l_x(\tau, x(\tau), u(\tau))$ and $b(\tau) = l_u(\tau, x(\tau), u(\tau))$. Then

$$
Dh(\xi) \cdot D^2 \mathcal{P}(\xi) \cdot (\zeta, \zeta) = \int_0^T a^T y + b^T w \, d\tau
$$

where $(y(\cdot), w(\cdot))$ satisfies (17). Equation (17) may be written as

$$
\dot{y} = (A(t) - B(t) K(t)) y(t) + D^2 f(x(t), u(t)) \cdot (\gamma_1(t), \gamma_2(t))
$$

where $A(t) = D_1 f(x(t), u(t))$ and $B(t) = D_2 f(x(t), u(t))$. Let $\Phi_c(\cdot, \cdot)$ be the state transition matrix of this differential equation. From the periodic condition, we have

$$
y(0) = [I - \Phi_c(T, 0)]^{-1} \int_0^T \Phi_c(T, \tau) D^2 f(x(\tau), u(\tau)) \cdot (\gamma_1(\tau), \gamma_2(\tau)) d\tau
$$

We have

$$
\begin{aligned}
& Dh(\xi) \cdot D^2 \mathcal{P}(\xi) \cdot (\zeta, \zeta) \\
&= \int_0^T a^T y + b^T w \, d\tau \\
&= \int_0^T \binom{a}{b}^T \binom{I}{-K} y \, d\tau \\
&= \int_0^T \binom{a}{b}^T \binom{I}{-K} \left[\Phi_c(\tau, 0) y(0) + \int_0^\tau \Phi_c(\tau, s) D^2 f(x(s), u(s)) \cdot (\gamma_1(s), \gamma_2(s)) ds \right] d\tau
\end{aligned}
$$

By change of order of integration, we obtain

$$
\begin{aligned}
& \int_0^T \int_0^\tau \Phi_c(\tau, s) D^2 f(x(s), u(s)) \cdot (\gamma_1(s), \gamma_2(s)) ds \, d\tau \\
&= \int_0^T \int_s^\tau \Phi_c(\tau, s) D^2 f(x(s), u(s)) \cdot (\gamma_1(s), \gamma_2(s)) d\tau \, ds
\end{aligned}
$$

Moreover,

$$\int_0^T \begin{pmatrix} a \\ b \end{pmatrix}^T \begin{pmatrix} I \\ -K \end{pmatrix} \Phi_c(\tau,0) y(0) d\tau$$

$$= \int_0^T \int_0^T \begin{pmatrix} a \\ b \end{pmatrix}^T \begin{pmatrix} I \\ -K \end{pmatrix} \Phi_c(\tau,0) [I - \Phi_c(T,0)]^{-1} \Phi_c(T,s) D^2 f(x(s),u(s))$$

$$\cdot (\gamma_1(s), \gamma_2(s)) d\tau ds$$

Combining we have,

$$Dh(\xi) \cdot D^2 P(\xi) \cdot (\zeta,\zeta) = \int_0^T q(s)^T D^2 f(x(s),u(s)) \cdot (\gamma_1(s), \gamma_2(s)) ds$$

where

$$q(t) = \Phi_c(T,t) q_T + \int_t^T \Phi_c^T(\tau,t)(I - K^T(\tau)) \begin{pmatrix} a(\tau) \\ b(\tau) \end{pmatrix} d\tau \tag{19a}$$

with

$$q_T = [I - \Phi_c(T,0)]^{-T} \int_0^T \Phi_c^T(\tau,0)(I - K^T(\tau)) \begin{pmatrix} a(\tau) \\ b(\tau) \end{pmatrix} d\tau \tag{19b}$$

Differentiating (19a) gives the differential equation

$$\dot{q}(t) = -A^T(t) q(t) - (I - K^T(t)) \begin{pmatrix} a(t) \\ b(t) \end{pmatrix} \tag{20a}$$

From (19b) follows the periodic condition

$$q(0) = q(T) \tag{20b}$$

Going back to the descent direction problem (18), it can be written as

$$\min_{(z(\cdot),v(\cdot))} \int_0^T \frac{1}{2} \begin{pmatrix} z(\tau) \\ v(\tau) \end{pmatrix}^T \begin{pmatrix} Q(\tau) & S(\tau) \\ R(\tau) & S^T(\tau) \end{pmatrix} \begin{pmatrix} z(\tau) \\ v(\tau) \end{pmatrix} + a^T(\tau) z(\tau) b^T(\tau) v(\tau) d\tau$$

$$\text{Subject to} \quad \dot{z} = A(t)z + B(t)v \tag{21}$$

$$\text{with} \quad z(0) = z(T)$$

where

$$\begin{pmatrix} Q(\tau) & S(\tau) \\ R(\tau) & S^T(\tau) \end{pmatrix} = \begin{pmatrix} l_{xx}(\tau,x(\tau),u(\tau)) & l_{xu}(\tau,x(\tau),u(\tau)) \\ l_{ux}(\tau,x(\tau),u(\tau)) & l_{uu}(\tau,x(\tau),u(\tau)) \end{pmatrix} + q(\tau)^T D^2 f(x(s),u(s))$$

Using the periodic condition and changing the order of integration etc. as we did in obtaining the q equation results in a periodic condition for the adjoint state. Then using the Riccati transformation, $p(t) = P(t)z(t) + r(t)$ leads to periodic conditions for $P(\cdot)$ and $r(\cdot)$.

7   SUMMARY AND CONCLUDING REMARKS

We outline a trajectory exploration approach to hybrid simulation, where we ask if there exist trajectories of the hybrid system (3) that are compatible with trajectories of the full system (2). For periodic trajectories, imposing the desired trajectories on the hybrid systems results in a TPBVP. If this exact imposition is not solvable, we may approach the problem in a least squares sense. Moreover in actual implementation, all the physical states are not measurable as assumed here, and states need to be estimated. The least squares problem and the experimental implementation of the trajectory exploration approach are subjects of current work.

ACKNOWLEDGEMENT

The authors gratefully acknowledge the financial support provided by MTS Systems Corp. for this work.

REFERENCES

Borutzky, W., Barnard, B. and Thoma, J.: 2002, An orifice flow model for laminar and turbulent conditions, *Simulation Modelling Practice And Theory* **10** (3–4), 141–152.

Hauser, J.: 2002, A projection operator approach to the optimization of trajectory functionals, *15th IFAC World Congress*, Barcellona.

Jelali, M. and Kroll, A.: 2003, *Hydraulic servo-systems: modelling, identification, and control*, Advances in industrial control, Springer, London; New York.

Merritt, H.E.: 1967, *Hydraulic control systems*, Wiley, New York.

CHAPTER 3

Integration schemes for real-time hybrid testing

P. Benson Shing
University of California, San Diego, USA

ABSTRACT: Real-time hybrid testing/simulation has gained much attention in recent years for studying the performance of structural systems and components as well as mechanical equipment under dynamic load conditions. It combines numerical simulation and physical testing in a complementary fashion so that the dynamic performance of a structural system or component can be evaluated in a realistic and efficient manner. This paper presents real-time hybrid test methods from a general perspective and an overview different explicit and implicit integration schemes that have been proposed for real-time tests. The implementation and pros and suitability of these schemes for real-time testing are discussed.

1 INTRODUCTION

In recent years, a number of real-time hybrid testing techniques have been proposed for studying the performance of structural systems and components as well as mechanical equipment under dynamic loads such as those induced by earthquake ground motions (Nakashima & Masaoka 1999; Horiuchi et al. 2000; Williams et al. 2001; Magonette 2001; Wallace et al. 2005; Bayer et al. 2005; Wu et al. 2006; Jung & Shing 2006; Bursi et al. in press). Hybrid testing is also often referred to as hybrid simulation. It combines numerical simulation and physical testing in a complementary fashion so that the dynamic performance of a structural system or component can be evaluated in a realistic and efficient manner. The term "hybrid" reflects the fact that part of a structure is analytically modeled while the remainder is physically tested. This technique is similar to classical substructure analysis methods. However, unlike a conventional substructuring approach, the partition of a structure into analytical and physical substructures in a real-time hybrid test can be based on the stiffness, damping, and inertial properties of the structure. For example, the inertial and damping properties of the physical substructure can be partially simulated analytically and partially represented in the physical specimen. For the extreme case that the dynamics of the physical specimen is entirely accounted for in the analytical model, one recovers the conventional pseudodynamic test method (Takanashi et al. 1975; Mahin & Shing 1985). Similar to the pseudodynamic test method, real-time hybrid testing relies on a numerical time integration scheme to solve of the equations of motion formulated for the analytical substructure.

Numerical integration schemes can be classified into implicit and explicit types. A key advantage of an implicit scheme is that it can be unconditionally stable, while explicit schemes are usually conditionally stable. However, an implicit scheme normally requires a Newton-type iterative method when the structure exhibits a nonlinear behavior (Jung 2005; Wei 2005; Jung et al. 2007). Explicit schemes were used for pseudodynamic tests in early years (Takanashi et al. 1975; Mahin & Shing 1985) as well as for real-time hybrid tests (Nakashima et al. 1999; Williams et al. 2001) because of their simplicity. Numerical schemes that are unconditionally stable and yet do not require an iterative correction have been proposed for non-real-time pseudodynamic tests, e.g., the OS method by Nakashima et al. (1990), the "explicit" scheme by Chang (2002), and the predictor-correct method by Bonelli and Bursi (2004). However, the use of these schemes for real-time hybrid testing has been limited. This paper presents real-time hybrid test methods from a general perspective and an overview of different integration schemes that have been proposed for such applications.

2   REAL-TIME HYBRID TEST METHODS

The basic concept of real-time hybrid testing is derived from the pseudodynamic test method which is based on the premise that the dynamic behavior of a structure can be accurately represented by a set of spatially discretized equations of motion.

$$\mathbf{Ma} + \mathbf{Cv} + \mathbf{r}_S = \mathbf{f} \tag{1}$$

in which $\mathbf{a}$ is the acceleration, $\mathbf{v}$ is the velocity, and $\mathbf{r}_S$ is the static restoring force vector of the structure; and $\mathbf{M}$ is the mass matrix, $\mathbf{C}$ is the viscous damping matrix, and $\mathbf{f}$ is the excitation force. The static restoring force vector $\mathbf{r}_S$ represents the forces that solely depend on the structural displacement $\mathbf{d}$ and its history. The general concept of hybrid testing is to partition a structure into two main portions with one represented by an analytical model and the other by a physical subassemblage or component that will be actually tested. The latter normally represents the portion of a structure that is difficult to model either because of its complicated structural nonlinearity or dynamics. For example, this can be a structural subassemblage that is most vulnerable to damage and that can develop an inelastic load-deformation behavior which is difficult to model analytically under the given loading condition. To convey this concept of structural partitioning, one can rewrite Equation 1 in the following form.

$$(\mathbf{M}^A + \mathbf{M}^E)\mathbf{a} + (\mathbf{C}^A + \mathbf{C}^E)\mathbf{v} + \mathbf{r}_S^A + \mathbf{r}_S^E = \mathbf{f} \tag{2}$$

in which the terms with superscript A represent the properties and response variables of the analytical model and the terms with superscript E are those of the physical component that is tested experimentally. Since both the inertia and damping forces developed by the physical component are measured quantities in a test, Equation 2 can be more appropriately expressed as

$$\mathbf{M}^A\mathbf{a} + \mathbf{C}^A\mathbf{v} + \mathbf{r}_S^A + \mathbf{r}^E = \mathbf{f} \tag{3}$$

with

$$\mathbf{r}^E = \mathbf{r}_I^E + \mathbf{r}_D^E + \mathbf{r}_S^E = \mathbf{M}^E\mathbf{a} + \mathbf{C}^E\mathbf{v} + \mathbf{r}_S^E \tag{4}$$

In the above equation, the damping forces $\mathbf{r}_D^E$ of the physical component do not necessarily have to be viscous in nature, and very often, it will be contributed by a number of different energy dissipation mechanisms such as Coulomb and hysteretic damping. Other forms of nonlinear damping mechanisms are also possible, especially if the physical subcomponent consists of special damping devices such as magneto-rheological dampers. However, velocity-independent damping forces can also be grouped into the term $\mathbf{r}_S^E$. Here, we adopt the definition that $\mathbf{r}_D^E$ represents velocity-dependent damping only.

In general, Equations 3 and 4 represent a variety of test methods. When $\mathbf{r}_I^E$ and $\mathbf{r}_D^E$ are close to zero, we recover the traditional pseudodynamic test method, including substructure testing, in which the physical substructure is loaded at a very slow rate. When all the terms with superscript A disappear, which is to say that no analytical model is assumed, we have the condition of the effective force test method (Dimig et al. 1999). In fact, it has been demonstrated that under this extreme, a real-time hybrid test can be conducted in the same way as an effective force test when a structural control strategy is combined with an implicit integration scheme to solve the governing equations of motion (Wu et al. 2007).

The different real-time test methods that have been proposed in the literature vary in the way the partitioned equation, i.e., Equation 3, is solved in the simulation process and in how the dynamic properties of the physical component are treated. As Equation 3 shows, in contrast to traditional pseudodynamic test methods, a real-time test does not require that the inertial and damping properties of the physical component be entirely represented by the analytical model. Hence, we can partition the terms in Equation 2 in the following way.

$$
\mathbf{a} = \left\{ \begin{array}{c} \mathbf{a}_A \\ \mathbf{a}_B \\ \mathbf{a}_E \end{array} \right\}
\mathbf{v} = \left\{ \begin{array}{c} \mathbf{v}_A \\ \mathbf{v}_B \\ \mathbf{v}_E \end{array} \right\}
\mathbf{r}_S^A = \left\{ \begin{array}{c} \mathbf{r}_{S,A} \\ \mathbf{r}_{S,B}^A \\ \mathbf{0} \end{array} \right\}
\mathbf{r}_S^E = \left\{ \begin{array}{c} \mathbf{0} \\ \mathbf{r}_{S,B}^E \\ \mathbf{r}_{S,E} \end{array} \right\}
\mathbf{f} = \left\{ \begin{array}{c} \mathbf{f}_A \\ \mathbf{f}_B^A + \mathbf{f}_B^E \\ \mathbf{f}_E \end{array} \right\}
\tag{5}
$$

$$
\mathbf{M}^A = \left[\begin{array}{ccc} \mathbf{M}_{AA}^A & \mathbf{M}_{AB}^A & \mathbf{0} \\ \mathbf{M}_{BA}^A & \mathbf{M}_{BB}^A & \mathbf{M}_{BE}^A \\ \mathbf{0} & \mathbf{M}_{EB}^A & \mathbf{M}_{EE}^A \end{array} \right]
\mathbf{M}^E = \left[\begin{array}{ccc} \mathbf{0} & \mathbf{0} & \mathbf{0} \\ \mathbf{0} & \mathbf{M}_{BB}^E & \mathbf{M}_{BE}^E \\ \mathbf{0} & \mathbf{M}_{EB}^E & \mathbf{M}_{EE}^E \end{array} \right]
\tag{6}
$$

in which subscripts A and E denote the degrees of freedom internal to the analytical and physical substructures, respectively, and B denotes those at the boundary of the two. It should be mentioned that in a real-time hybrid test, the mass of the physical specimen can be less than that of the actual structural model being considered due to the fact that the inertial effect can be simulated analytically. For this reason, the mass associated with the degrees of freedom of the physical substructure can be divided into an analytical part, with superscript A, and a physical part, with superscript E. For the sake of generality, we can partition the damping matrix in the same fashion as the mass matrix.

Different explicit and implicit integration schemes can be used to solve Equation 3 numerically. The selection of a particular integration scheme or schemes depends to a certain extent on the properties of the analytical and physical substructures and how the two are coupled in the analysis. One approach that is referred to as the *total approach* is discussed here. In this approach, it is assumed that the dynamics of the entire system can be accurately described by the degrees of

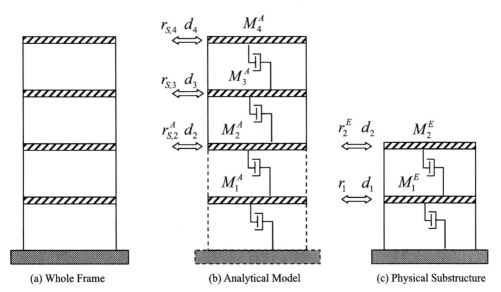

| (a) Whole Frame | (b) Analytical Model | (c) Physical Substructure |

Figure 1. Partitioning of a structure into analytical and physical substructures with a total approach.

freedom identified in Equation 3. Such a case is illustrated with the example shown in Figure 1, for which Equation 3 can be expressed as follows.

$$
\begin{bmatrix} M_1^A & 0 & 0 & 0 \\ 0 & M_2^A & 0 & 0 \\ 0 & 0 & M_3^A & 0 \\ 0 & 0 & 0 & M_4^A \end{bmatrix} \begin{Bmatrix} a_1 \\ a_2 \\ a_3 \\ a_4 \end{Bmatrix} + \begin{bmatrix} C_1^A + C_2^A & -C_2^A & 0 & 0 \\ -C_2^A & C_2^A + C_3^A & -C_3^A & 0 \\ 0 & -C_3^A & C_3^A + C_4^A & -C_4^A \\ 0 & 0 & -C_4^A & C_4^A \end{bmatrix} \begin{Bmatrix} v_1 \\ v_2 \\ v_3 \\ v_4 \end{Bmatrix}
$$

$$
+ \begin{Bmatrix} 0 \\ r_{S,2}^A \\ r_{S,3} \\ r_{S,4} \end{Bmatrix} + \begin{Bmatrix} r_1 \\ r_2^E \\ 0 \\ 0 \end{Bmatrix} = \begin{Bmatrix} f_1 \\ f_2 \\ f_3 \\ f_4 \end{Bmatrix} \tag{7}
$$

in which a damping coefficient C_i^A represents a dash-pot in story i. This approach is identical to a standard analysis procedure with the physical substructure treated as a *super element* whose resistance is assembled into the global force vector of the structure.

3 INTEGRATION SCHEMES

3.1 *Implicit schemes with iteration*

Many implicit integration schemes are unconditionally stable. This is certainly an attractive feature when one has to deal with many degrees of freedom and high frequency modes in the analytical model. Nevertheless, implicit schemes normally require an iterative solution procedure for nonlinear structures. This poses a major challenge for real-time hybrid testing as the actuators controlling structural displacements have to move continuously with a smooth velocity, and the displacement commands to the actuators have to be sent at the sampling frequency of the digital controller. However, with a standard Newton-type iterative solution scheme, the displacement corrections will be progressively reduced as the trial displacements approach a converged solution. Hence, updating the displacement commands directly with the trial displacements will lead to undesired velocity fluctuations from one time step to the next. In addition, the number of iterative corrections required in each time step will vary depending on the degree of nonlinearity developed in the structure. This uncertainty is not acceptable for a real-time test where a converged solution has to be attained within a designated time interval.

To circumvent the aforementioned problem, a special iterative method that adopts a fixed number of iterations can be used. This method will be presented here using the implicit Newmark integration method as an example. With Equation 3, the Newmark method can be expressed as follows.

$$
\mathbf{M}^A \mathbf{a}_{i+1} + \mathbf{C}^A \mathbf{v}_{i+1} + \mathbf{r}_{S,i+1}^A + \mathbf{r}_{i+1}^E = \mathbf{f}_{i+1} \tag{8}
$$

$$
\mathbf{d}_{i+1} = \tilde{\mathbf{d}}_{i+1} + \Delta t^2 \beta \mathbf{a}_{i+1} \tag{9}
$$

$$
\mathbf{v}_{i+1} = \tilde{\mathbf{v}}_{i+1} + \Delta t \gamma \mathbf{a}_{i+1} \tag{10}
$$

where

$$
\tilde{\mathbf{d}}_{i+1} = \mathbf{d}_i + \Delta t \mathbf{v}_i + \Delta t^2 (1/2 - \beta) \mathbf{a}_i \tag{11}
$$

$$
\tilde{\mathbf{v}}_{i+1} = \mathbf{v}_i + \Delta t (1 - \gamma) \mathbf{a}_i \tag{12}
$$

with β and γ being the numerical parameters that determine the specific integration method. With $\beta = 1/4$ and $\gamma = 1/2$, we have the constant-average-acceleration (trapezoidal) method.

In an iterative solution, if $\mathbf{d}_{i+1}^{(k)}$ is the trial solution in iteration step k and $\mathbf{r}_{S,i+1}^{A(k)}$ and $\mathbf{r}_{i+1}^{E(k)}$ are the associated forces obtained from the analytical and physical substructures, respectively, the equilibrium error can be calculated with Equations 8 through 10 as follows.

$$\mathbf{R}_{i+1}^{(k)} = \overline{\mathbf{M}}(\tilde{\mathbf{d}}_{i+1} - \mathbf{d}_{i+1}^{(k)}) - \mathbf{C}^A \tilde{\mathbf{v}}_{i+1} - \mathbf{r}_{S,i+1}^{A(k)} - \mathbf{r}_{i+1}^{E(k)} + \mathbf{f}_{i+1} \tag{13}$$

where

$$\overline{\mathbf{M}} = \frac{\mathbf{M}^A + \Delta t \gamma \mathbf{C}^A}{\Delta t^2 \beta} \tag{14}$$

Assuming that damping in the physical substructure is viscous, we can have the following modified Newton iteration scheme based on the initial stiffness of the structure to arrive at a corrected displacement vector.

$$\mathbf{K}^* \Delta \mathbf{d}_{i+1}^{(k)} = \mathbf{R}_{i+1}^{(k)} \tag{15}$$

where the effective stiffness matrix is defined as

$$\mathbf{K}^* = \frac{(\mathbf{M}^A + \mathbf{M}^E) + \Delta t \gamma (\mathbf{C}^A + \mathbf{C}^E)}{\Delta t^2 \beta} + (\mathbf{K}_0^A + \mathbf{K}_0^E) \tag{16}$$

The updated displacement is then obtained as

$$\mathbf{d}_{i+1}^{(k+1)} = \mathbf{d}_{i+1}^{(k)} + \Delta \mathbf{d}_{i+1}^{(k)} \tag{17}$$

This scheme requires that we accurately estimate the values of $\mathbf{M}^E$, $\mathbf{C}^E$, and $\mathbf{K}_0^E$. However, since they are only used to estimate the updated displacements in the iteration process, their exact values are not needed even though this may affect the convergence rate. In fact, it is desirable that their values are a little (few percent) higher than the exact ones to avoid overshoot and to preserve the stability of the numerical solution (Shing et al. 1991a, 1991b).

As mentioned previously, for a real-time hybrid test, one must avoid undesired velocity fluctuations from one iteration step to the next and attain convergence within a fixed time interval. To achieve this, a special correction procedure that has a fixed number of iterations in each time step is presented here (Jung and Shing 2007). As shown in Figure 2, this method relies on an interpolation technique to assure a smooth motion of an actuator during iteration. Each iteration consists of two steps: (1) the trial displacement $d_{i+1}^{(k+1)}$ for each degree of freedom is evaluated as in a conventional Newton-type iterative process; and (2) a fraction of the trial displacement, which is termed the desired displacement $d_{i+1}^{\mathrm{d}(k+1)}$, is then computed with a quadratic interpolation and imposed on the structural specimen. Except for the first time step, three points are used for the construction of the interpolation function: the trial displacement $d_{i+1}^{(k+1)}$ in the current iteration, and the displacements $d_{i-1}^{\mathrm{d}(n)}$ and $d_i^{\mathrm{d}(n)}$, which are computed in the very last iteration of the two previous time steps, with n denoting the total number of iterations specified in each time step and k varying from 0 to $(n-1)$. In the first time step, the interpolation will be based on the initial velocity and displacement.

Once iteration is completed in a time step, the response needs to be updated so that the computation for the next time step can begin. Since the number of iterations in each time step is fixed, convergence errors are expected to be larger than those with a conventional Newton-type iteration. In addition, convergence errors can be exacerbated by the time delays in actuator response and in the data exchange between the controller/data-acquisition processor and the target computer.

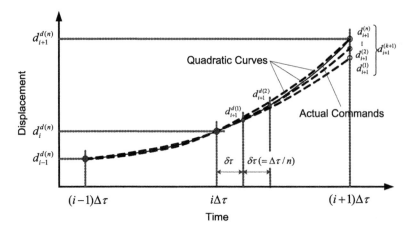

Figure 2.   Iteration scheme.

To minimize the effect of these errors and enforce equilibrium at the end of each time step, the following approximate correction can be introduced for updating the displacements and forces in the last iteration.

$$\mathbf{d}_{i+1} = \mathbf{d}_{i+1}^{d(n)} \tag{18}$$

$$\mathbf{r}_{S,i+1}^{A} = \mathbf{r}_{S,i+1}^{A(n-1)} + \mathbf{K}_0^{A}[\mathbf{d}_{i+1}^{d(n)} - \mathbf{d}_{i+1}^{(n-1)}] \tag{19}$$

$$\mathbf{r}_{i+1}^{E} = \mathbf{r}_{i+1}^{m(n-1)} + \mathbf{K}^{E*}[\mathbf{d}_{i+1}^{d(n)} - \mathbf{d}_{i+1}^{m(n-1)}] \tag{20}$$

in which $\mathbf{d}_{i+1}$, $\mathbf{r}_{S,i+1}^{A}$, and $\mathbf{r}_{i+1}^{E}$ are the updated displacements and forces that are treated as the converged solutions carried over to the next time step, $\mathbf{d}_{i+1}^{d(n)}$ is the desired displacement vector computed in the n-th iteration, which is the last iteration, $\mathbf{d}_{i+1}^{m(n-1)}$ and $\mathbf{r}_{i+1}^{m(n-1)}$ are the displacements and restoring forces measured from the test structure at the beginning of the last iteration, and $\mathbf{K}^{E*}$ is the portion of the effective stiffness formulated in Equation 16 that contains only the properties of the physical substructure. With $\mathbf{d}_{i+1}$ computed, the velocities and accelerations can then be updated with Equations 9 and 10 and the computation is advanced to the next time step.

Another issue to address is that during each iteration, the displacements imposed on the physical substructure will be different from the desired displacements computed by the interpolation method. To avoid displacement incompatibility between the physical and analytical substructures at the boundary degrees of freedom, it is desired that the displacements and forces measured from the physical substructure be corrected in each iteration as follows (Wei 2005).

$$\mathbf{d}_{i+1}^{(k)} = \mathbf{d}_{i+1}^{d(k)} \tag{21}$$

$$\mathbf{r}_{i+1}^{E(k)} = \mathbf{r}_{i+1}^{m(k)} + \mathbf{K}^{E*}[\mathbf{d}_{i+1}^{d(k)} - \mathbf{d}_{i+1}^{m(k)}] \tag{22}$$

The iteration scheme presented here can be used with other implicit integration methods. Shing et al. (2006) has applied this scheme with the α-method (Hughes 1983). However, the α-method presents a consistency issue when the inertia forces developed by the physical substructure become significant and cannot be ignored.

3.2   *Explicit schemes*

One of the most commonly used explicit integration methods in structural dynamics is the central difference method. This method has been used by a number of researchers for non-real-time pseudodynamic testing as well as real-time hybrid tests because of its simplicity and computational efficiency. In this method, the structural velocity and acceleration are approximated by the central difference equations.

$$\mathbf{v}_i = \frac{\mathbf{d}_{i+1} - \mathbf{d}_{i-1}}{2\Delta t} \tag{23}$$

$$\mathbf{a}_i = \frac{\mathbf{d}_{i+1} - 2\mathbf{d}_i + \mathbf{d}_{i-1}}{\Delta t^2} \tag{24}$$

Substituting the above velocity and acceleration approximations into the equations of motion, i.e., Equation 3, at time step i, one can solve for $\mathbf{d}_{i+1}$ as follows.

$$\mathbf{d}_{i+1} = \left(\mathbf{M}^A + \frac{\mathbf{C}^A \Delta t}{2}\right)^{-1} \left[\Delta t^2 (\mathbf{f}_i - \mathbf{r}_{S,i}^A - \mathbf{r}_i^E) + 2\mathbf{M}^A \mathbf{d}_i - \left(\mathbf{M}^A - \frac{\mathbf{C}^A \Delta t}{2}\right)\mathbf{d}_{i-1}\right] \tag{25}$$

in which $\mathbf{r}_i^E$ is measured from the physical substructure in each time step. It is a two-step method. To calculate the displacements $\mathbf{d}_1$ in the first step, one needs to know the values of $\mathbf{d}_{-1}$ which is a fictitious quantity that can be obtained from Equations 23 and 24 as follows.

$$\mathbf{d}_{-1} = \mathbf{d}_0 - \Delta t \, \mathbf{v}_0 + \frac{\Delta t^2}{2} \mathbf{a}_0 \tag{26}$$

in which $\mathbf{a}_0$ is calculated from the equilibrium condition by knowing initial conditions of the structure.

$$\mathbf{a}_0 = \frac{1}{\mathbf{M}^A} (\mathbf{f}_0 - \mathbf{C}^A \mathbf{v}_0 - \mathbf{r}_{S,0}^A - r_0^E) \tag{27}$$

In spite of the fact that this scheme is computationally efficient, it is conditionally stable. Furthermore, to compute the displacement commands for the next time step, a small time duration is needed. On the other hand, for digital actuator controllers, displacement commands have to be received at each sampling interval of the controller, which is normally much smaller than the integration time interval Δt that is needed. To assure that the digital controller will receive displacement commands without interruption, different implementation schemes have been proposed. One is a staggered approach (Nakashima et al. 1992) in that the displacements for step $i + 1$ are computed by using the displacements obtained at steps $i - 1$ and $i - 3$, whereas the displacements for step $i + 2$ are computed with those at steps i and $i - 1$. With this scheme, the integration time interval will be $2\Delta t$ instead of Δt with Δt equal to the sampling time interval of the controller.

Another approach is to use an interpolation and extrapolation strategy (Nakashima et al. 1999). In this approach, there is a "Signal Generation Task" to generate displacement signals by extrapolating the displacements obtained in previous time steps, and the extrapolated signals are sequentially sent to the controller with an interval of δt, which is the sampling time interval of digital controller. During that time, a "Response Analysis Task" is executed to solve the equations of motion with an integration time interval Δt to compute the response in the next time step. When the displacements for the next time step are computed, the "Signal Generation Task" stops extrapolation and starts interpolation to reach the new displacement target Figure 3 illustrates this procedure.

To evaluate the accuracy of this procedure, a series of real-time hybrid tests were conducted by Nakashima et al. (1999) for single- and multi-degree-of-freedom structures. They have found that

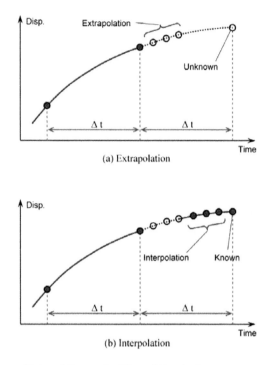

(a) Extrapolation

(b) Interpolation

Figure 3.   Extrapolation and interpolation method for real-time testing.

the extrapolation scheme is able to ensure stable and accurate results up to the response frequency of 3.0 Hz if the number of extrapolations is not greater than six. In addition, the order of the polynomial function for extrapolation should be higher than three to obtain reliable results.

As a final note, the central difference method is mathematically identical to the Newmark method (Equations 8 through 12) with $\beta = 0$ and $\gamma = 0.5$. However, the explicit form of the Newmark method is superior to the central difference method in terms of the numerical conditioning. Furthermore, because of the inevitability of actuator control errors, the calculated displacements should be used instead of the measured values for the numerical computation using Equation 25 (Shing & Mahin, 1985), and the forces measured should be corrected in the same way as that shown in Equation 20.

3.3   Unconditionally stable schemes with no iteration

Several unconditionally stable schemes that do not require iteration have been proposed for non-real-time pseudodynamic testing. They are referred to as either implicit or explicit schemes. However, these schemes are based on implicit formulations, and iteration is avoided by performing either a prediction or correction that is based on the initial stiffness properties of the structure. Hence, strictly speaking, these schemes should be considered as either predictor-corrector or implicit methods.

The Operator Splitting (OS) method was first adopted by Nakashima et al. (1990) for non-real-time pseudodynamic tests. It is based on an explicit prediction-implicit correction approach. To extend this method for real-time tests, the equations of motion are expressed as follows.

$$\mathbf{M}^A \mathbf{a}_{i+1} + \mathbf{C}^A \mathbf{v}_{i+1} + (\mathbf{K}^A + \mathbf{K}^E)(\mathbf{d}_{i+1} - \hat{\mathbf{d}}_{i+1}) + \mathbf{C}^E(\mathbf{v}_{i+1} - \hat{\mathbf{v}}_{i+1})$$
$$+ \mathbf{M}^A(\mathbf{a}_{i+1} - \hat{\mathbf{a}}_{i+1}) + \hat{\mathbf{r}}^A_{S,i+1} + \hat{\mathbf{r}}^E_{i+1} = \mathbf{f}_{i+1} \tag{28}$$

in which $\hat{\mathbf{d}}_{i+1} = \tilde{\mathbf{d}}_{i+1}$ is the predictor displacement vector calculated with Equation 11, $\hat{\mathbf{v}}_{i+1}$ and $\hat{\mathbf{a}}_{i+1}$ are the associated velocity and acceleration vectors, and $\hat{\mathbf{r}}^A_{S,i+1}$ and $\hat{\mathbf{r}}^E_{i+1}$ are the corresponding forces. In this method, the predictor displacements $\hat{\mathbf{d}}_{i+1}$ are imposed on the physical substructure with velocities $\hat{\mathbf{v}}_{i+1}$ and accelerations $\hat{\mathbf{a}}_{i+1}$ in each time step. The forces $\hat{\mathbf{r}}^E_{i+1}$ developed by the physical substructure are then measured, while $\hat{\mathbf{r}}^A_{S,i+1}$ will be computed. With these, $\mathbf{d}_{i+1}$, $\mathbf{v}_{i+1}$, and $\mathbf{a}_{i+1}$ can be calculated with Equations 28, and 9 through 12. However, the main difficulty of this method for real-time tests is the calculation and imposition of the predictor velocities and accelerations, $\hat{\mathbf{v}}_{i+1}$ and $\hat{\mathbf{a}}_{i+1}$, which are not provided by the integration scheme. Wu et al. (2005) have applied this scheme to test a damper system where the actual mass of the specimen is negligible. They assume that $\hat{\mathbf{v}}_{i+1}$ is constant in each time step with $\hat{\mathbf{a}}_{i+1}$ implied to be zero. They have concluded that by having the prediction velocity that is not consistent with that of the OS method, the unconditional stability of the integration scheme can no longer be preserved.

The unconditionally stable "explicit" scheme of Chang (2002) adopts an explicit prediction that is based on the initial elastic stiffness of a structure. For a linearly elastic structure, this method is essentially identical to the constant-average-acceleration method. However, its accuracy for highly nonlinear structural behavior has not been thoroughly studied. This also applies to the Rosenbrock algorithms proposed by Bursi et al. (in press). The Rosenbrock algorithms do not require iteration but the multi-stage algorithms require the computation of structural responses at intermediate steps.

The use of these methods for real-time tests faces the same challenge as other implicit and explicit methods in that displacement commands have to be issued at the sampling frequency of the digital controller. For this, the staggered and extrapolation—interpolation procedures as proposed for the explicit schemes can also be used here.

4 CONCLUSIONS

This paper presents real-time hybrid test methods from a general perspective and an overview different explicit and implicit integration schemes that have been used for real-time tests. Each type of schemes has its merits as well as drawbacks. Unconditionally stable implicit schemes that require iterative correction are computational less efficient but their accuracy for highly nonlinear systems have been well proven. Explicit schemes are computationally efficient. However, they are conditionally stable, which presents a problem when high-frequency modes are present in the analytical substructure. Unconditionally stable implicit schemes that do not require iteration have been recently extended for real-time tests. However, their accuracy for highly nonlinear systems has not been well studied. The technical difficulties associated with the implementation of some of these schemes for real-time tests have not been completely resolved. In spite of these limitations, real-time hybrid tests have been proven to be a promising technique that combines numerical simulation and physical testing in a complementary manner.

ACKNOWLEDGMENTS

Some of the material presented in this paper is based on the previous research of the author supported by the National Science Foundation under the Cooperative Agreement No. 0086592. However, opinions expressed in this paper are those of the authors and do not necessarily represent those of the sponsor.

REFERENCES

Bayer, V., Dorka, U.E., Füllekrug, U., & Gschwilm, J. 2005. On real-time pseudodyanmic sub-structure testing: algorithm, numerical and experimental results. *Aerospace Science and Technology* 9: 223–232.

Bonelli, A. & Bursi, O.S. 2004. Generalized-α methods for seismic structural testing. *Earthquake Engineering and Structural Dynamics* 33: 1067–1102.

Bursi, O.S., Gonzalez-Buelga, A., Vulcan, L., Neild, S.A., & Wagg, D.J. (in press). Novel coupling Rosenbrock-based algorithms for real-time dynamic substructure testing. *Earthquake Engineering & Structural Dynamics*.

Chang, S.Y. 2002. Explicit pseudodynamic algorithm with unconditional stability. *Journal of Engineering Mechanics* 128(9): 935–947.

Dimig, J., Shield, C., French, C., Bailey, F., & Clark, A. 1999. Effective force testing: a method of seismic simulation for structural testing. *J. Struct. Eng.* 125(9): 1028–1037.

Horiuchi, T., Masahiko, I., & Konno, T. 2000. Development of a real-time hybrid experimental system using a shaking table. *Proc. 12th World Conference on Earthquake Engineering*: Paper No. 0843. Auckland, New Zealand.

Hughes, T.J.R. 1983. Analysis for transient algorithms with particular reference to stability behavior. In T. Belyschko & T.J.R Hughes (eds.), *Computational Methods for Transient Analysis*. Amsterdam: North-Holland.

Jung, R.-Y., 2005. Development of real-time hybrid test system. *Doctoral Thesis*, University of Colorado at Boulder, Boulder, Colorado.

Jung, R.-Y. & Shing, P.B. 2006. "Performance evaluation of a real-time pseudodynamic test system. *Earthquake Engineering and Structural Dynamics* 35(7): 789–810.

Jung, R.-Y. & Shing, P.B. 2007. Performance of a real-time pseudodynamic test system considering nonlinear structural response. *Earthquake Engineering and Structural Dynamics* 36(12): 1785–1809.

Magonette, G. 2001. Development and application of large-scale continuous pseudodynamic testing techniques. *Phil. Trans. Royal Society* 359: 1771–1799.

Mahin, S.A. & Shing, P.B. 1985. Pseudodynamic method for seismic testing. *Journal of Structural Engineering* 111: 1482–1503.

Nakashima, M., Kaminosono, T., & Ishida, M. 1990. Integration techniques for substructure pseudodynamic test. *Proc. 4th US National Conference on Earthquake Engineering* II: 515–524.

Nakashima, M., Kato, H., & Takaoka, E. 1992. Development of real-time pseudodynamic testing. *Earthquake Engineering and Structural Dynamics* 21: 79–92.

Nakashima, M. & Masaoka, N. 1999. Real-time on-line test for MDOF systems. *Earthquake Engineering and Structural Dynamics* 28: 393–420.

Shing, P.B. & Mahin, S.A. 1985. Computational aspects of a seismic performance test method using on-line computer control. *Earthquake Engineering and Structural Dynamics* 13: 507–526.

Shing, P.B., Vannan, M.T., & Carter, E. 1991. Implicit time integration for pseudodynamic tests. *Earthquake Engineering and Structural Dynamics* 20(6): 551–576.

Shing, P.B. & Vannan, M.T. 1991. Implicit time integration for pseudodynamic tests: convergence and energy dissipation. *Earthquake Engineering and Structural Dynamics* 20(6): 809–819.

Shing, P.B., Stavridis, A., Wei, Z., Stauffer, E., Wallen, R., & Jung, R. 2006. Validation of a fast hybrid test system with substructure tests. *Proc. 17th Analysis and Computation Conference*, SEI/ASCE, St. Louis, Missouri.

Takanashi, K. et al. 1975. Non-linear earthquake response analysis of structures by a computer actuator on-line system, part 1—details of the system. *Trans. of Architectural Inst. of Japan* 229: 77–83.

Wallace, M.I., Sieber, J., Neild, S.A., Wagg, D.J., & Krauskopf. 2005. Stability analysis of real-time dynamic substructuring using delay differential equation models. *Earthquake Engineering and Structural Dynamics* 34(15): 1817–1832.

Wei, Z. 2005. Fast hybrid test system for substructure evaluation. *Doctoral Thesis*, University of Colorado at Boulder, Boulder, Colorado.

Williams, D.M., Williams, M.S., & Blakeborough, A. 2001. Numerical modeling of a servo-hydraulic testing system for structures. *Journal of Engineering Mechanics* 127(8): 816–827.

Wu, B., Xu, G., Wang, Q., & Williams, M.S. 2006. Operator-splitting method for real-time substructure testing. *Earthquake Engineering and Structural Dynamics* 35(3): 293–314.

Wu, B., Wang, Q., Shing, P.B., & Ou, J. 2007. Equivalent force control method for generalized real-time substructure testing with implicit integration. *Earthquake Engineering and Structural Dynamics* 35(9): 1127–1149.

CHAPTER 4

Assessment of experimental errors in hybrid simulation of seismic structural response

G. Mosqueda
University at Buffalo, Buffalo, New York, USA

T.Y. Yang & B. Stojadinovic
University of California, Berkeley, California, USA

ABSTRACT: Hybrid simulation outcomes are sensitive to measurement and control errors associated with the experimental substructures. These errors must be successfully mitigated in order to obtain reliable simulation results. To asses the severity of experimental errors during a hybrid simulation, an on-line error monitoring method is presented that predicts the reliability of the test results during the experiments. Quantitative estimates of error are derived based on energy added to the hybrid model due to measurable errors in the experimental substructures. A recently completed hybrid simulation of seismic response of an innovative lateral load resisting system, a suspended zipper braced frame is used to demonstrate the effectiveness of the proposed on-line error monitoring indicators. It is shown that a well-calibrated simulation conducted using state-of-the-art equipment can deliver accurate results.

1 INTRODUCTION

Hybrid simulation is an experimentally based method for investigating the response of structure to dynamic excitation using a hybrid model. A hybrid model is an assemblage of one or more physical and one or more numerical, consistently scaled, substructures. The equation of motion of a hybrid model under dynamic excitation is solved during a hybrid simulation test. Experiments have shown that results from hybrid simulation and shaking table tests are comparable, but only when propagation of experimental errors is successfully mitigated in the hybrid simulation (Takanashi and Nakashima, 1987; Mahin et al., 1989; Magonette and Negro, 1998). Errors are introduced into hybrid simulations through the structural model idealization, the approximate numerical integration methods used to solve the equation of motion, and the experimental setup. The results of a hybrid simulation are most sensitive to experimental errors, because these errors are not known prior to testing and can be large in improperly tuned experimental setups.

In an effort to reduce experimental errors, researchers have focused on reducing actuator tracking errors (difference in command displacement and measured displacement) through better instrumentation for feedback measurement and improvements of the servo-hydraulic actuator control loop (Takanashi and Nakashima, 1987; Thewalt and Mahin, 1987; Stoten and Magonette, 2001). Actuator tracking errors can be measured during a hybrid simulation and this information has been used to correct the measured restoring force based on the initial stiffness of the structure (Shing et al., 2002) or by adjusting the time increment of the simulation step to accommodate the measured displacement (Yi and Peek, 1993). To mitigate the response lag of the actuators, researchers have proposed constant (Horiouchi et al., 1999, Reinhorn et al., 2004) and variable (Darby et al., 2002, Ahmadizadeh et al., 2007) compensation schemes. Even though the magnitude of experimental errors has been reduced using methods in these previous studies, the errors cannot be completely eliminated. Relatively small systematic experimental errors can have a cumulative effect on the simulation result that can compromise the stability of executing the simulation, particularly for high frequency modes, and bring into question the validity of the simulation (Mosqueda et al., 2007b).

Hybrid Simulation Error Monitors (*HSEM*) can be used to predict the quality of the hybrid simulation results and detect unacceptable levels of systematic experimental errors in real-time (Mosqueda et al., 2007a). *HSEM* are based on normalized measures of cumulative energy added to the hybrid simulation as a result of actuator tracking errors. They can provide instantaneous information on the control state of each actuator or the cumulative effect of the actuators control errors on the overall accuracy of the computed structural response. *HSEM* is applied here to a hybrid simulation of the seismic response of a suspended zipper braced frame lateral load resisting system to demonstrate the accuracy and reliability of the results and to demonstrate the minimal propagation of errors in modern hybrid simulations using state-of-the-art hardware to control the experimental substructures. The suspended zipper braced frame was selected because accurate simulation of complex interaction of element- and system-level nonlinear response experienced by this lateral load system is very sensitive to experimental and actuator tracking errors.

2   EXPERIMENTAL ERRORS

Errors are introduced into hybrid simulations through the structural model idealization, the approximate numerical integration methods used to solve the equation of motion, and the experimental setup. Modeling and numerical errors, resulting primarily from the assumptions in the analytical modeling techniques, are described by Shing and Mahin (1983, 1984). Thewalt and Mahin (1987) provide a thorough discussion of hardware components and sources of experimental errors. Of these various sources of errors, experimental errors can have the most substantial impact on the simulation results, mostly because these errors are not known prior to testing and can be large for improperly tuned experimental setups.

Experimental errors result from the displacement control of hydraulic actuators, force relaxation or strain rate effects due to the slow rates of testing, calibration errors in the instrumentation, and noise generated in the instrumentation and analog to digital converters. Experimental errors can be classified as either random or systematic in nature. A major source of random errors is low amplitude, high frequency noise in experimental measurements, which can excite spurious response in lightly damped higher modes in the hybrid model. To mitigate this effect, higher modes can be damped numerically by using specific integration algorithms (Hilbert et al., 1977; Shing and Mahin, 1984) or by solving the governing equation of motion in its integral form (Chang et al., 1999).

Systematic errors can have a greater influence on the simulation results compared to random errors (Thewalt and Roman, 1994; Mosqueda et al., 2005). Sources of systematic errors include load-history effects on the experimental substructures and displacement control errors in the servo-hydraulic actuators. For example, in a conventional ramp-and-hold pseudo-dynamic test, the specimen may exhibit force-relaxation during the hold portions of the test. Continuous (Magonette, 2001; Stojadinovic et al., 2006) and real-time (Nakashima, 2001) testing techniques successfully mitigate this source of errors in pseudo-dynamic tests. Further, real-time testing methods can be used to load the experimental substructures at realistic seismic rates for strain-rate sensitive materials. Alternatively, strain-rate effects have been compensated mathematically during slow continuous tests (Molina et al., 2002b).

Actuator tracking errors are also composed of random and systematic components. In a hybrid simulation, the numerical integrator observes the experimental substructure behavior in terms of command displacement and measured reaction force at each controlled degree of freedom. However, a more accurate representation of the structural specimen behavior is given by the measured displacement and the measured force. This difference between the measured and command displacement modifies the observed structural behavior and may alter the energy that is added to the numerical simulation as a result of the errors in the experimental setup. For example, the actuator response lag can have a negative damping effect, but it can be compensated using the polynomial based feed-forward method proposed by Horiuchi et al. (1999) to minimize this effect.

Effective procedures have been proposed to mitigate the propagation of errors in a hybrid simulation, mainly through improved servo-hydraulic control. However, there is no guarantee that these methods will work effectively throughout the test. A procedure is discussed next that measures the severity of the errors as the test progresses. The proposed error measures can function as the basis for a real-time monitoring system to asses the accuracy of an on-going hybrid simulation.

3   ENERGY BALANCE FOR HYBRID SIMULATION

The hybrid simulation test method comprises numerical simulations and simultaneous experimental testing of substructure components by integrating the dynamic equation of motion for the hybrid model. A time-stepping integration procedure is used to solve the discretized equation of motion for displacements, $\mathbf{u}$, at the degrees of freedom of the structure at time intervals $t_i = i \, \Delta t$ for $i = 1 \cdots N$.

$$\mathbf{M}\ddot{\mathbf{u}}_i + \mathbf{C}\dot{\mathbf{u}}_i + \mathbf{r}_i + \mathbf{r}_i^E = \mathbf{f}_i \tag{1}$$

The subscript i denotes the time-dependant variables at time t_i, Δt is the integration time step in the numerical simulation and N is the number of integration steps. The mass matrix, $\mathbf{M}$, damping matrix, $\mathbf{C}$, and applied loading, $\mathbf{f}$, are typically modeled numerically in the computer. For the hybrid model, the restoring force vector is assembled using forces measured at the degrees of freedom of experimental substructures, $\mathbf{r}_i^E$, or computed for numerical substructures, $\mathbf{r}_i$.

The energy balance equation is obtained by determining the work done by each of the force components in Eq. (1). Integrating with respect to the displacement $\mathbf{u}$, and using relative displacements formulation results in (Uang and Bertero, 1990):

$$\int (\mathbf{M}\ddot{\mathbf{u}})^T \, d\mathbf{u} + \int (\mathbf{C}\dot{\mathbf{u}})^T \, d\mathbf{u} + \int (\mathbf{r})^T \, d\mathbf{u} + \int (\mathbf{r}^E)^T \, d\mathbf{u} = \int (\mathbf{f})^T \, d\mathbf{u} \tag{2}$$

In addition to kinetic energy and viscous damping dissipation, energy can be stored and dissipated by both numerical ($\mathbf{r}$) and experimental ($\mathbf{r}^E$) substructures. The energy associated with the experimental substructures is of particular interest: this energy corresponds to the work done by the measured resisting forces on the displacements of the experimental substructure.

$$E^E = \int (\mathbf{r}^E)^T \, d\mathbf{u} = \int (\mathbf{r}^m + \mathbf{r}^{error})^T \, d\mathbf{u} \tag{3}$$

In Eq. (3), $\mathbf{r}^m$ = measured force and $\mathbf{r}^{error}$ = presumed error in the force measurements, which are then introduced into the energy balance equation. Substituting Eq. (3) and moving the error term to the right-hand side in Eq. (2) results in

$$\int (\mathbf{M}\ddot{\mathbf{u}})^T \, d\mathbf{u} + \int (\mathbf{C}\dot{\mathbf{u}})^T \, d\mathbf{u} + \int (\mathbf{r})^T \, d\mathbf{u} + \int (\mathbf{r}^m)^T \, d\mathbf{u} = \int (\mathbf{f})^T \, d\mathbf{u} - E^{error} \tag{4}$$

The negative sign on the experimental substructure energy error, E^{error}, indicates that a negative energy error adds energy to the structural model, similar to the effect of negative damping.

3.1   *Source of energy errors*

In a typical experimental setup for hybrid simulation, the displacements are imposed at the actuator degrees of freedom and the forces are measured by load cells on the actuators. Therefore, it is useful to express the energy in the experimental substructures in terms of the actuator degrees of freedom coordinates. This approach allows for the energy errors to be computed separately for each actuator and related to the energy in the global structural model using similitude scaling factors and geometric transformations from conventional finite element models.

The geometric transformation matrix, $\mathbf{T}$, is obtained by transforming the global structural degrees of freedom, $\mathbf{u}$, to the actuator degrees of freedom, $\mathbf{u}^{ac} = \mathbf{T}\mathbf{u}$, where $\mathbf{u}^{ac}$ is a vector containing the displacement command signals for the actuators. The transformation $\mathbf{r}^m = \mathbf{T}^T\mathbf{r}^{am}$, can be used to transform the measured forces at the actuator degrees of freedom, $\mathbf{r}^{am}$, to the global degrees of freedom in the restoring force vector $\mathbf{r}^m$. Substituting the transformations and their derivatives into Eq. 3, an expression of the experimental substructures energy in terms of the actuator degrees of freedom is obtained.

$$\mathbf{E}^{\mathrm{E}} = \int (\mathbf{r}^{\mathrm{am}})^{\mathrm{T}} \mathbf{T} \, d\mathbf{u} \tag{5}$$

In the above expression, the vector multiplication suggests that energy in the experimental substructures is simply the sum of energy contributions produced by each individual actuator. Further, Eq. (5) states that the numerical integration algorithm considers the response of the experimental elements as the measured restoring forces resulting from the command displacements. Since the command displacement is not necessarily the same as the applied or measured displacement, the behavior of the specimen may not be correctly modeled.

A more accurate representation of the energy in experimental substructures can be obtained from the measured displacement and the measured force data. This data pair is the best representation of the experimental substructure available and is typically used to evaluate a structure after a quasi-static test. The best estimate of energy stored and dissipated by the experimental sub-structure is

$$\mathbf{E}^{\mathrm{BE}} = \int (\mathbf{r}^{\mathrm{am}})^{\mathrm{T}} \, d\mathbf{u}^{\mathrm{am}} \tag{6}$$

where $\mathbf{u}^{\mathrm{am}} = $ measured displacement of the specimen at the actuator degrees of freedom. It is important to consider that in a hybrid simulation, the load path of the experimental element should coincide with the computed response history of the structural model. Consequently, $\mathbf{E}^{\mathrm{BE}}$ is not necessarily the best energy estimate with respect to numerical simulation since the load history on the experimental substructure may differ from that in the numerical model. In spite of this, the energy error introduced into a hybrid simulation can be estimated as

$$\mathbf{E}^{\mathrm{error}} = \mathbf{E}^{\mathrm{BE}} - \mathbf{E}^{\mathrm{E}} \tag{7}$$

Eq. 7 shows the difference between the energy that is dissipated by the experimental substructures and the energy dissipated by the numerical integration algorithm. The energy error, $\mathbf{E}^{\mathrm{error}}$, will be close to zero if there is no displacement control errors.

It is important to note that Eq. (7) does not account for scaling factors between the numerical model (global degrees of freedom) and the experimental substructure (actuator degrees of freedom). Such scaling factors, if not included in the transformation matrix, $\mathbf{T}$, should be included in the energy calculations to provide the correct comparison between the computed energy errors and the relevant energy measure in Eq. (2).

3.2   *Hybrid simulation error monitors*

The experimental energy error estimate presented above can be computed in each step of the numerical integration algorithm. Further, the error estimate only requires information from the current and previous steps. Thus, energy errors can be computed throughout the simulation and used to monitor their severity in real-time. The use of *HSEM* to asses the quality of a hybrid simulation requires that a threshold value be specified as the level of unacceptable error, corresponding to the point when the accuracy of simulation results is exceeding an allowable tolerance limits. To this end, a normalization of the energy error by the input energy from the earthquake excitation is considered here. A global view of the experimental substructure energy errors is given by the energy balance Eq. (4), where the input energy may be defined as

$$\mathbf{E}^{input} = \int (\mathbf{f})^{T} \, d\mathbf{u} \tag{8}$$

A measure of energy error relative to the input energy is defined as follows:

$$HSEM = \frac{E^{error}}{E^{input} + E^{strain}} \tag{9}$$

The strain energy, E^{strain} is included in the denominator because the input energy is zero at the beginning of a simulation, which can result in an infinite value for $HSEM \cdot E^{strain}$ is defined here as the maximum recoverable strain energy assuming the structural behavior is elastoplastic. Note that for an elastic system, the energy input from the earthquake is dissipated only by viscous damping. Therefore, in the linear range, a comparison of the energy error to the energy dissipated by viscous damping will provide similar results.

4   LARGE-SCALE TEST OF A SUSPENDED ZIPPER BRACED FRAME

Suspended zipper braced frame is an innovative steel concentrically braced frame proposed by Leon and Yang (2003). The structural system has very similar configuration to that of the conventional inverted-V braced frame, except for a zipper column, a structural element added between the beam mid-span points from the second to the top stories of the frame. In the event of severe earthquake shaking, the inverted-V braces are designed to buckle to dissipate earthquake energy. This creates unbalanced vertical forces at the mid span of the beams. The zipper columns are designed to transfer the unbalanced vertical forces to the higher stories by engage the remaining unbuckled braces and beams to resist the unbalanced vertical force. If the earthquake shaking increases, brace buckling will propagate to the higher stories until all compressed braces buckle. To prevent the system from entering a force-deformation response softening range, the top-story braces are designed to remain linearly-elastic under the total unbalanced vertical forces in the system.

The response of the suspended zipper braced frame is complex, with interacting nonlinear response of the braces and nonlinear response of the system. Hybrid simulation was selected to examine the seismic response of this structural system because it provides insight into the minutia of the local buckling response of the braces simultaneously with the global view of the system response at every instant of the earthquake loading. The hybrid simulation was conducted at the newly developed *nees@berkeley* laboratory, an equipment site of the George E. Brown Jr. Network for Earthquake Engineering Simulation (http://nees.berkeley.edu). New, state-of-the-art high performance actuators, controllers and data acquisition system were used in this simulation, providing a unique opportunity to evaluate the effectiveness of the proposed *HSEM*.

The hybrid model of the suspended zipper braced frame consists of a physical first-story inverted-V braced sub-assembly, set up in the laboratory, and a finite element model of the remainder of the frame (Figure 1). The substructures were selected to capture the complex brace buckling behavior experimentally since they are difficult to model analytically, while the remainder of the structure is modeled using a state-of-the-art finite element software, OpenSees (UCB, 1997). The computer model consists of flexibility-formulation nonlinear beam-column elements (de Souza, 2000; Filippou and Fenves, 2004) with fiber sections and zero-length elements to model the beams, braces, zipper columns and columns of the suspended zipper braced frame. The foundations of

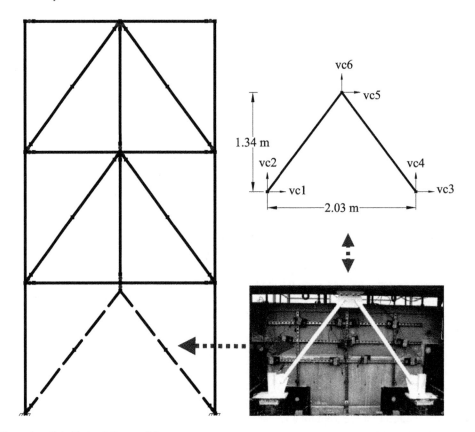

Figure 1.   Hybrid simulation model.

the columns and the beam-to-column connections were modeled using the semi-rigid connection model suggested by Astaneh-Asl (2005). Gravity load was ignored in the analytical simulation. Table 1 shows the element sizes of the tested 1/3-scale suspended zipper braced frame model. The reduced scale size of the experimental model was governed by collaborative earthquake simulator testing of an identical model at University at Buffalo (Schachter and Reinhorn, 2006). The floor heights of the 1/3-scale suspended zipper braced frame are 1.34 m (52.75 in.), 1.29 m (50.75 in.) and 1.29 m (50.75 in.) for the first, second and third story, respectively. The bay width is 2.03 m (80 in.). A floor mass of 9063 kg (621 slugs) was assigned as two lumped masses at the exterior nodes at each floor. Rayleigh mass and stiffness proportional damping of 5 percent was assigned to the first and second vibration modes of the hybrid model.

The ground motion used in this study, LA22, was selected from the suite of ground motions used in the SAC Joint Venture project (Somerville et al., 1997) for buildings located in Los Angeles. The ground motion was selected and scaled according to the procedure presented in Schachter and Reinhorn (2006) for the comparative earthquake simulator studies of a similar frame.

Figure 2 shows the test setup for the suspended zipper braced frame hybrid simulation at the *nees@berkeley* laboratory. The setup consists of two vertical dynamic actuators, one horizontal dynamic actuator, and a guiding system to allow the intersection of the inverted-V braces to only move in-plane. Nonlinear geometry transformation of the actuator movement is accounted for in this hybrid simulation test setup (Yang, 2006).

This hybrid simulation used a displacement controlled algorithm to solve the system dynamics. The integrator (Newmark average-acceleration time-step integration) implemented in OpenSees calculated the structural deformation due to external excitation at the beginning of each time step.

Table 1.   Element sizes for the 1/3-scale suspended zipper braced frame model.

Story	Braces	Column	Beam	Zipper column
3	HSS $3 \times 3 \times 3/16$	S4 × 9.5	S3 × 5.7	HSS $2 \times 2 \times 3/16$
2	HSS $2 \times 2 \times 1/8$	S4 × 9.5	S5 × 10	HSS $1.25 \times 1.25 \times 3/16$
1	HSS $2 \times 2 \times 1/8$	S4 × 9.5	S3 × 7.5	

Figure 2.   Test setup for the suspended zipper braced frame hybrid simulation at the *nees@berkeley* laboratory.

The test setup then executes the displacement and samples the force feedback from the experimental subassembly. Simultaneously, the analytical elements in OpenSees calculate the force feedback at the displacement state. The integrator then combines the force feedback from the experimental and the analytical elements and the external excitation to calculate the structural deformations at the next time step. Detailed procedures for the hybrid simulation algorithm and architecture are presented in Yang (2006).

5   EXPERIMENTAL RESULTS AND ASSESSMENT OF ERRORS

The response of the structure applied at the horizontal actuator degree of freedom is shown in Figure 3. Both, the displacement command and feedback displacement measurements corresponding to the horizontal actuator are shown to demonstrate the performance of the actuator controller. Due to careful tuning of the system prior to testing, actuator tracking errors are negligible, and thus the experimental errors are expected to be acceptably small.

The proposed *HSEM* are used to better assess the effects of experimental errors on the response of the hybrid model in this simulation. Figure 4 shows the energy stored as strain energy or dissipated by hysteric damping in the experimental substructure. These plots include the energy input in the substructure from all three actuators and compare the energy estimates using Eqs. 5 and 6. The difference between these two curves is the estimated energy error according to Eq. 7, which is shown in Figure 5. The error is negative, which indicates that energy is being added to the

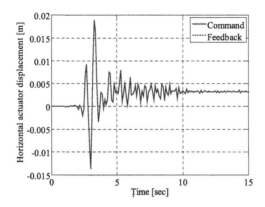

Figure 3.   Displacement command and feedback histories of the horizontal actuator.

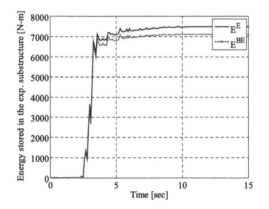

Figure 4.   Energy stored and dissipated by experimental substructure.

experiment as a result of the actuator tracking errors. Note that the energy error is approximately 5% of the energy input by the actuators into the experimental substructure.

The effects of energy error on the overall results of the hybrid simulation are best demonstrated through the proposed *HSEM*. Figure 6 presents the *HSEM* described in Eq. 9, which is the energy error normalized by the total input earthquake energy plus the strain energy. For this purpose, the strain energy is calculated as

$$E^{strain} = \frac{1}{2} \mathbf{u}_\mathbf{y}^t \, \mathbf{K} \mathbf{u}_\mathbf{y} \tag{10}$$

where $\mathbf{u}_\mathbf{y}$ = a vector of yield displacement and $\mathbf{K}$ = initial stiffness matrix. In this study the yield displacement is estimated to be 0.2% of the inter-story drift ratio. The results indicate that the energy error resulting from the experimental setup is less than 2% of the total energy input to the structural model from the external excitation. This error occurred during the approximately 3-second long acceleration spike in the ground motion record shown in Figure 3, the remainder of the response computation did not contribute significantly to the *HSEM*. The error of this magnitude is reasonably small. It indicates that the performance of the experimental setup is very good and the final results are accurate within an acceptable margin. As previously discussed, sources of errors in a hybrid simulation, other than actuator tracking and control should also be assessed to ensure simulation results are correct.

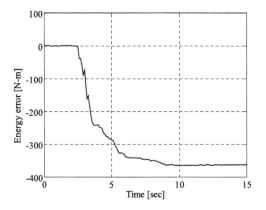

Figure 5.   Energy error in the experimental substructure.

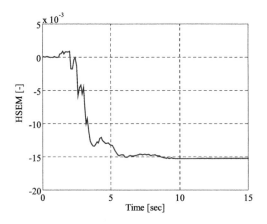

Figure 6.   *HSEM* (Equation 9).

6   CONCLUSIONS

Hybrid simulation outcomes are sensitive to measurement and control errors associated with the experimental substructures. Hybrid simulation error monitors (*HSEM*) are proposed in this paper to asses the severity of experimental errors during a hybrid simulation. They are based on quantifying energy added to the hybrid model due to measurable command errors in the experimental substructures. A recently completed hybrid simulation of seismic response of an innovative lateral load resisting system, a suspended zipper braced frame is used to demonstrate the effectiveness of the proposed on-line error monitoring indicators. This demonstration was successful: *HSEM* can be easily implemented and used to measure the experimentally induced simulation error (primarily due to actuator control). It was shown the suspended zipper braced frame seismic response simulation experienced an addition of approximately 5% of the total excitation energy due to actuation errors. This error level was considered acceptable. More important, use of state-of-the-art controllers and actuators, together with accurate calibration and exacting test setup preparation is key to conducting highly accurate hybrid simulations, even when the response of the hybrid model is highly nonlinear and quite complex. Ongoing work is addressing development of additional measures to assess the quality of a hybrid simulation.

REFERENCES

Ahmadizadeh, M., Mosqueda, G. and Reinhorn, A.M. (2007). "Compensation of actuator delay and dynamics for real-time hybrid structural simulation." *Earthquake Eng. Struct. Dyn.*, in press.

Astaneh-Asl, A. (2005). "Design of shear tab connections for gravity and seismic loads." *SteelTips*, Structural Steel Educational Council.

Chang, S.-Y., Tsai, K.-C. and Chen, K.-C. (1999). "Improved time integration for pseudodynamic tests." *Earthquake Eng. Struct. Dyn.*, 27(7), 711–730.

Darby, A.P., Williams, M.S. and Blakeborough, A. (2002). "Stability and delay compensation for real-time substructure testing." *J. Eng. Mech.*, 128(12), 1276–1284.

de Souza, R.M. 2000. "Force-based finite element for large displacement inelastic analysis of frames." Doctoral dissertation, *Graduate Division of the University of California at Berkeley*, Univ. of California at Berkeley, Berkeley, Calif.

Filippou, F.C. and Fenves, G.L. (2004). "Methods of analysis for earthquake-resistant structure." Chapter 6 in Earthquake Engineering From Engineering Seismology to Performance-based Engineering, *RCR Press*, Boca Raton, Florida.

Hilbert, H.M., Hughes, T.J.R. and Taylor, R.L. (1977). "Improved numerical dissipation for time integration algorithms in structural dynamics." *Earthquake Eng. Struct. Dyn.*, 5, 283–292.

Horiuchi, T., Inoue, M., Konno, T. and Namita, Y. (1999). "Real-time hybrid experimental system with actuator delay compensation and its application to a piping system with energy absorber." *Earthquake Eng. Struct. Dyn.*, 28(10), 1121–1141.

Leon, T. Roberto and Yang, C.S. 2003. "Special Inverted-V-braced Frames with Suspended Zipper Struts." International Workshop on Steel and Concrete Composite Construction, *IWSCCC*, National Center for Research on Earthquake, Taipei, Taiwan.

Magonette, G.E. and Negro, P. (1998). "Verification of the pseudodynamic test method." *European Earthquake Eng.*, XII(1), 40–50.

Magonette, G. (2001). "Development and application of large-scale continuous pseudo-dynamic testing techniques." *Phil. Trans. R. Soc. Lond.*, 359, 1771–1799.

Mahin, S.A., Shing, P.B., Thewalt, C.R. and Hanson, R.D. (1989). "Pseudodynamic test method—Current status and future direction." *J. Struct. Eng.*, 115(8), 2113–2128.

Molina, F.G., Verzeletti, G., Magonette, G., Buchet, P.H., Renda, M., Geradin, M., Parducci, A., Mezzi, M., Pacchiarotti, A., Federici, L., and Mascelloni, S. (2002). "Pseudodynamic tests on rubber base isolators with numerical substructuring of the superstructure and strain-rate effect compensation." *Earthquake Eng. Struct. Dyn.*, 31(98), 1563–1582.

Mosqueda, G., Stojadinovic, B. and Mahin, S.A. (2007a). "Real-time error monitoring for hybrid simulation. I: *Methodology* and experimental verification." *J. Struct. Eng.*, 133(8), 1100–1108.

Mosqueda, G., Stojadinovic, B. and Mahin, S.A. (2007b). "Real-time error monitoring for hybrid *simulation*. I: Structural response modification due to errors." *J. Struct. Eng.*, 133(8), 1109–1119.

Nakashima, M. (2001). "Development, potential, and limitations of real-time online (pseudo-dynamic) testing." *Phil. Trans. R. Soc. Lond.*, 359, 1851–1867.

Newmark, N.M. (1959). "A method of computation for structural dynamics." *J. Eng. Mech.*, 85(3), 67–94.

Schachter, M., Reinhorn, A. and Leon, R. (2006). "On the importance of 3D analysis in zipper frames." *8th National Conference on Earthquake Engineering*, Paper 1621, San Francisco, CA, April 2006.

Shing, P.B. and Mahin, S.A. (1983). "Experimental error propagation in pseudodynamic testing." *Report UCB/EERC-83/12* Earthquake Eng. Res. Center, Univ. of California at Berkeley, Berkeley, Calif.

Shing, P.B. and Mahin, S.A. (1984). "Pseudodynamic test method for seismic performance evaluation: theory and implementation." *Report UCB/EERC-84/01*. Earthquake Eng. Res. Center, Univ. of California at Berkeley, Berkeley, Calif.

Shing, P.B., Spacone, E., and Stauffer, E. (2002). "Conceptual design of fast hybrid test system ad the University of Colorado." *Proc., 7th U.S. National Conf. Earthquake Engineering*, Earthquake Engineering Research Institute (EERI), Boston.

Somerville, P., Smith, N., Punyamurthula, S. and Sun, J. (1997). "Development of ground motion time history for phase 2 of the FEMA SAC steel project." *SAX/BD-97/04*, SAC Joint Venture, Sacramento, CA, USA.

Stojadinovic, B., Mosqueda, G. and Mahin, S.A. (2006). "Event-driven control system for geographically distributed hybrid simulation." *ASCE, J. Struct. Eng.* 132(1), 68–77.

Stoten, D., and Magonette, G. (2001). "Developments in the Automatic Control of Experimental Facilities." *ECOEST2/ICONS Report No. 9*, Laboratoria Nacional de Engenharia Civil, Lisboa, Portugal.

Takanashi, K. and Nakashima, M. (1987). "Japanese activities on on-line testing." *J. Eng. Mech.*, 113(7), 1014–1032.

Thewalt, C.R. and Mahin, S.A. (1987). "Hybrid solution techniques for generalized pseudodynamic testing." *Rep. No. UCB/EERC-87/09*, Earthquake Eng. Res. Center, Univ. of California at Berkeley, Berkeley, Calif.

Thewalt, C.R. and Roman, M. (1994). "Performance parameters for pseudodynamic tests." *J. Struct. Eng.*, 120(9), 2768–2781.

Uang, C.-H. and Bertero, V.V. (1990). "Evaluation of seismic energy in structures." *Earthquake Eng. Struct. Dyn.*, 19(1), 77–90.

UCB (1997). "Open system for earthquake engineering simulation (opensees) framework", *Pacific Earthquake Engineering Research Center*, University of California, Berkeley, http://opensees.berkeley.edu/.

Yang, T.Y. (2006). "Performance evaluation of innovative steel braced frames." Doctoral dissertation, *Graduate Division of the University of California at Berkeley*, Univ. of California at Berkeley, Berkeley, Calif.

Yi, W.H. and Peek, R. (1993). "Posterior time-step adjustment in pseudodynamic testing." *J. Eng. Mech.*, 119(7), 1376–1386.

CHAPTER 5

Hard versus soft real time hybrid simulations: An experimental assessment

E. Stauffer
DYS Services, Denver, Colorado, USA

G. Haussmann
University of Colorado, Boulder, Colorado, USA

K. Smith
University of Colorado, Boulder, Colorado, USA

ABSTRACT: The University of Colorado (CU) Fast Hybrid Testing (FHT) facility utilizes an unconditionally stable implicit time integration technique to combine numerical modeling and an experimental substructure into a unified hard realtime earthquake simulation. Originally the FHT system was designed and developed with the understanding that the experimental test component would behave as a structural element, which is to say that it would react to an imposed displacement history at its boundary with a corresponding force history. System identification test data indicates a damping device is best represented as a component which reacts to an imposed velocity history at its boundary with a corresponding force history. This understanding has made it necessary to generalize the formulation of the implicit time integration algorithm so as to correctly accommodate the velocity based nature of a nonlinear damping device. Results utilizing the new generalized formulation are presented here.

1   INTRODUCTION

Recent testing at the CU NEES FHT facility has focused on highly nonlinear damping devices, Magneto-Rheological (MR) dampers, that have been tested to evaluate there potential as semi-active earthquake damage mitigation devices, Christenson et al. (2005). A series of additional, internally conducted, damper tests have focused on issues of FHT system verification and validation. In order to complete these tests the direct time integration algorithm used at CU NEES was generalized (Stauffer, 2006), allowing for the participation of an experimental test element with highly nonlinear damping properties.

Realtime (RT) pseudo-dynamic or hybrid simulations necessitate that there is a one to one correspondence between the prototype time, simulation time and testing time. In a RT hybrid simulation some efficiency and cost savings is achieved by dividing the test structure into a physical component (the damper) and an analytical component (the structure) creating a hybrid model. Only the physical substructure needs to be constructed and tested in the laboratory while the remaining analytical substructure is modeled within a computer, using finite element methods.

Several terms that will be used throughout this article will be defined to provide a clearer understanding of their use as applied to hybrid simulations. Hybrid models may be divided into two distinct conceptual components. One will be physically tested in the laboratory and will be designated the *experimental component*, while the remaining portion of a model will reside in some form of mathematical representation and will be designated the *numerical component*. These two components combine to makeup the *hybrid model* or structure.

Returning to the notion of time it will be helpful to define three additional terms involving time. *Prototype time* corresponds to passage of time during a recorded event such as the El Centro earthquake on May 18, 1940, in the Imperial Valley of California. *Simulation time* is the marking of time within a computer simulation and may or may not relate coherently to the passage of time as marked by a clock on the wall. Typical direct time integration techniques establish equilibrium at discrete intervals of *simulation time* that are Δt seconds apart. *Testing time* is the marking of the passage of time during, in our case, a hybrid simulation and can be thought of as the clock on the laboratory wall. Given a consistent starting point and that all three definitions of time are equivalent over the widest possible range of time intervals, we have hard realtime conditions.

2   A LINEAR DAMPER MODEL FOR ALGORITHM VALIDATION

Phenomenological models of the MR damper have been presented by many researchers among them Spencer et al. (1997) and Gavin (2001). By restricting the maximum velocity of the input motion the complex nonlinear behavior of the MR damper is avoided and the simplified model presented here provides good accuracy. As Figure 1 illustrates the force vs. velocity curves generated by the velocity limited MR damper system ID tests have a distinctly elliptical shape.

A linear two term model can be used to accurately replicate this behavior.

$$f_d = c_d \dot{x} + k_d x \tag{1}$$

where the parameters c_d and k_d are selected to provide a response that matches the system ID data. The motion for the system ID tests is harmonic which implies that Equation 1 may be rewritten as

$$f_d = b_1 \cos(\omega t) + b_2 \sin(\omega t) \tag{2}$$

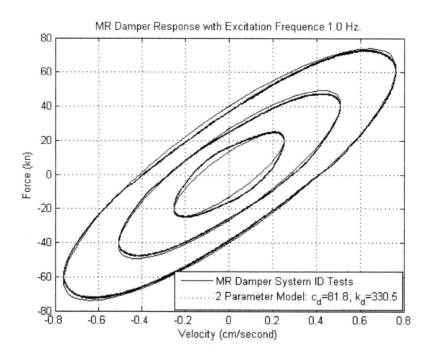

Figure 1.   MR damper and 2 parameter model response curves.

where

$$b_1 = c_d x_0 \omega \tag{3}$$

$$b_2 = k_d x_0 \tag{4}$$

The values of x_0 and ω correspond to the displacement amplitude and frequency of excitation respectively. A summary of the values used to generate subsequent fully numerical simulation results involving the two parameter model are provided in Table 1.

Figure 1 shows a comparison of the results obtained from the two parameter damper model and the data collected during the system ID testing. At this frequency the model provides a good match of the damper's measured response, especially in the higher amplitude range.

3   A SIMPLE HYBRID MODEL

For validation purposes a single column structure will be tested that consists of a single steel W 14 × 257 column which is fully fixed at its base. A concentrated mass is added to the top of the column. The MR damper is added as a horizontal element connected at the top of the column. The single column structure is illustrated in Figure 2.

The FHT results obtained from this simple model are compared directly to accurate fully numerical solutions. The fully numerical solution uses the 2 parameter damper model described previously in place of the MR damper. When the relative velocity of the two end points or nodes at each end of the MR damper are appropriately limited ($v_{max} \leq 0.762$ cm./second) the force-velocity response for the MR damper is similar to, but not exactly, that of the idealized linear viscous damper model.

The algorithm accuracy tests were all conducted using a scaled horizontal base impulse function for excitation with the scaling done to achieve the velocity limitation described above. A total of 12 different impulse response tests were completed.

Table 1.   Linear damper model parameters.

Excitation frequency	x_0 (cm)	v_0 (cm/sec)	b_1	b_2	c_d	k_d
$\omega = 2\pi$	0.1212	0.762	62.3	40.1	81.8	330.5
$\omega = \pi$	0.2426	0.762	83.15	28.0	109.1	115.5

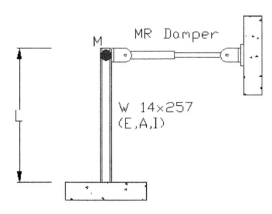

Figure 2.   Arrangement of single column hybrid model.

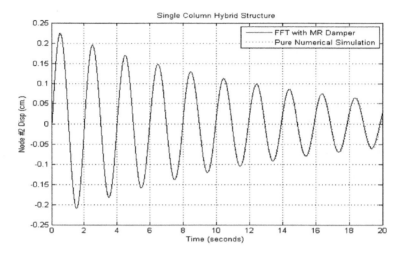

Figure 3.   Impulse response comparison $\omega = \pi$ and $\zeta = 0.02$.

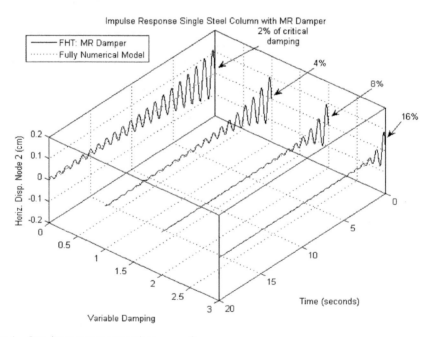

Figure 4.   Impulse response comparison $\omega = \pi 2$.

For these tests consider the representation of the governing system of equations expressed in uncoupled or modal form where there are i modes and the equation of motion for each mode may be expressed as

$$\ddot{x}_i + 2\omega_i \zeta_i \dot{x}_i + \omega_i x_i = f_i \quad (i = 1, 2, 3) \tag{5}$$

To investigate the accuracy of the FHT integration procedure under a variety of conditions the values of ω and ζ for the first bending mode will be varied and tested for the corresponding hybrid structure. In order to do this the length of the column and the discrete mass at the top of the column will be calculated for three different frequencies ($\omega = 2\pi, \pi, 0.2\pi$) and 4 different damping values ($\zeta = 0.02, 0.04, 0.08, 0.16$). The results from these hard RT hybrid simulations are compared to accurate purely numerical simulations providing a basis for evaluating the accuracy of the FHT scheme. For the pure numerical simulations a 2 parameter linear viscous damping model is used in place of the MR damper.

The results shown in Figure 3 are typical of the series of twelve base impulse response tests and indicate the high level of accuracy provided by the hybrid integration algorithm. A small amount of frequency distortion is apparent which may be attributed to a combination of imperfection in the 2 parameter damper model used for the pure numerical simulation and the presently unimplemented force correction, which compensates for actuator command-response error.

Four additional impulse response comparisons are provided for the case involving the single column structure with a fundamental frequency of $\omega = 2\pi$ in Figure 4.

The quality of matching between the FHT and the purely numerical simulation begins to degrade slightly as the amplitude of oscillatory motion drops well below 0.1 cm. This can be attributed to two influences. First, imperfections in the 2 parameter damper model that become more apparent at lower amplitudes (see Figure 1). Secondly, a reduction in the signal to noise ratio for both the displacement and the force signals that feedback to the hybrid integration algorithm.

In summary it is understood that these are demanding and revealing tests. A good level of accuracy has been demonstrated over a reasonably broad range of the possibilities with the acknowledgement that improvement is always needed and desirable.

4   IMPORTANCE OF REALTIME IN DAMPER BASED HYBRID SIMULATION

An additional series of tests were completed to investigate the effects of compromising the condition of hard RT during hybrid simulations involving the MR damper. By distorting the scaling of time in a variety of ways during a hybrid simulation the importance of correct time scaling is explored. Five different types of time distortions were selected to represent a variety of compromised testing conditions.

Two consistently scaled time tests were carried out with a consistent and uniform expansion of time. This was done using the same program and hardware as the hard RT simulations with minor modifications allowing for the consistent expansion of testing time. As would be expected when the time scaling factor is reduced, from 100 to 10, the solution approaches the hard RT result. The distortion or expansion of simulation time is uniform and consistent throughout the entire test.

A varying time distortion in the form of a random delay between 200 and 400 milliseconds that occurs at only 0.2 percent of the computation cycles is also shown. For the remaining computation cycles (99.8 percent) hard realtime conditions are satisfied.

Two additional cases for explicit time integration are also considered, and their time histories are shown in Figure 5. Both involve the use of OpenFresco—the first using a series of rapid ramp and holds and the second using extrapolation and interpolation. The extrapolation/interpolation approach provides command signal continuity (Mosqueda, 2005) and was developed at the University of California Berkeley. In the initial phase of the integration time step, during the computation of the next solution point, a command signal is generated that extrapolates forward in time based on prior solutions. Once the solution is available a transition to interpolation takes place. An event driven approach is used to implement this that accommodates a total of five states (extrapolate, interpolate, slow, hold, and free-vibration). OpenSees is used to carry out the computations using the α-operator split integrator on one computer, running under a non-realtime operating system, and the interpolation/extrapolation command generation is carried out on a second computer in a realtime computing environment.

Figures 5 and 6 shows the improvement achieved by moving from a ramp and hold approach to the extrapolation and interpolation implemented in OpenFresco. For comparison purposes a Hybrid Component Compliance (HCC) factor is defined. It is based on the two accurate solutions derived from implicit time integration. The HCC is calculated at each discrete instance of time throughout

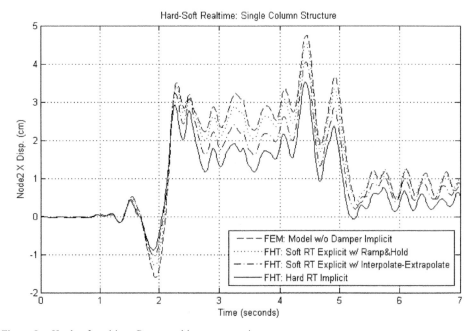

Figure 5. Hard-soft realtime: Response history comparison.

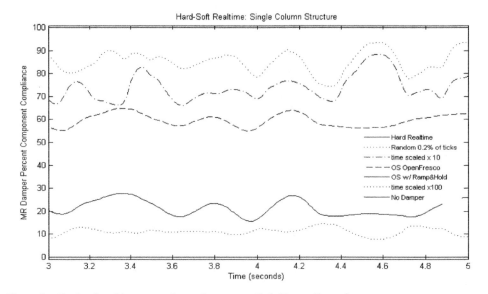

Figure 6. Hard-soft realtime—experimental component hybrid compliance factor.

the simulation using

$$HCC_i = \frac{X_i^{SimSoft} - X_i^{NoDamper}}{|X_i^{HRT} - X_i^{NoDamper}|} \times 100 \qquad (6)$$

where the X_i^{HRT} is the ith solution for the implicit hard realtime case which fully engages the hybrid experimental component, $X_i^{NoDamper}$ is the ith value of the implicit solution that has no

Table 2. Mean value of HCC during El Centro strong motion.

Simulation case	Mean HCC
0.2% random	85.7
OS OpenFresco	59.4
Consistently scaled ($\times 10$)	74.4
OS Ramp & Hold	21.1
Consistently scaled ($\times 100$)	11.0

Mean values calculated for $3 < t < 5$ only.

hybrid experimental component and $X_i^{SimSoft}$ is the ith solution for the simulation involving some alteration of testing time deeming it a soft realtime simulation. A HCC value of 100 indicates full compliance and a solution that matches exactly the accurate (hard realtime) implicit solution involving the damper. A value of 0 indicates a total lack of compliance and a solution that matches exactly the damper free simulation, certainly an undesirable condition with regard to simulation error.

During the strong motion portion of the simulation, $(3 < t_{sim} < 5)$, the HCC is plotted in Figure 5 and graphically indicates that the solution with 0.2 percent random time distortions provides the best level of accuracy of all the time distorted simulations. The mean value of the HCC for the various soft realtime cases is summarized in Table 2.

For the simple single column structure the least distorted of the soft realtime cases studied here is the intermittent random distortion with a mean HCC of approximately 86. By most measures this is not an adequate level of accuracy and it is difficult to envision a research and testing scenario for which this level of accuracy is acceptable. It is also difficult to establish a priori that a given relaxation of hard realtime conditions will yield a result with some sufficient level of accuracy.

5 CONCLUSION

The CU NEES FHT laboratory is a hybrid testing facility with strict hard realtime simulation capabilities. Experience with the unconditionally stable implicit direct time integration algorithm used at CU has confirmed the robustness and accuracy of this approach. Rate sensitive materials and devices such as the MR damper necessitate that hybrid simulations be carried out in hard realtime. The significant effects shown in this report of relatively minor time distortions on MR damper hybrid simulations highlight the need for a careful and accurate representation of time in the simulated event. One might be tempted to argue that there exists a gray area within which limited time distortions lead to relatively minor and acceptable errors. As researchers familiar with nonlinear systems know, there exists a very fine sensitivity of these systems to initial conditions and every effort must be taken to reduce sources of uncertainty and error. For researchers and engineers developing and testing new earthquake resistant materials, devices or designs, especially ones with demonstrated rate sensitivities, it is no longer necessary to work and conduct research in this gray area.

ACKNOWLEDGEMENT

The authors gratefully acknowledge the financial support of NEES Inc. and the National Science Foundation. The cooperation and assistance of the CU NEES FHT laboratory and its staff is also acknowledged.

REFERENCES

Christenson, R.E. and Emmons, A.T., 2005, "Semiactive structural control of a nonlinear building model: Considering reliability," *ASCE Structures Congress 2005: Session on Semiactive Control of Civil Structures*, New York City, New York.

Gavin, H.P., 2001, Multi-Duct ER Dampers, *Journal of Intelligent Material Systems and Structures*, Vol. 12, No. 5, pp. 353–366.

Hilber, H.M., Hughes, T.J.R., and Taylor, R.L., 1977, Improved numerical dissipation for time integration algorithms in structural dynamics, Earthquake eng. Struct. Dyn. 5, 283–292.

Hughes, T.J.R., 1983, Analysis for transient algorithms with particular reference to stability behavior, Belyschko T, Hughes TJR, Editors. *Computational methods for transient analysis*. Amsterdam, North-Holland.

Mosqueda, G., Stojadinovic, B., and Mahin, S., 2005, Implementation and accuracy of continuous hybrid simulation with geographically distributed substructures. Report No. EERC 2005-02 Earthquake Engineering Research Center, College of Engineering, University of California, Berkeley. November 2005.

Shing, P.B., Spacone, E. and Stauffer, E., 2002, Conceptual design of fast hybrid test system at the University of Colorado. Proceeding, Seventh U.S. National Conference of Earthquake Engineering, Boston.

Shing, P.B., Vannan, M.T., and Carter, E., 1991, Implicit time integration for pseudodynamic tests, Earth. Engrg, and Struct. Dynamics, Vol. 20, 551–576.

Spencer, B.F., Jr., Dyke, S.J., Sain, M.K., and Carlson, J.D. (1997). Phenomenological Model of a Magnetorheological Damper, *Journal of Engineering Mechanics, ASCE*, Vol. 123, No. 3, pp. 230–238.

Stauffer, E., 2006, The CU-Boulder Fast Hybrid Test: Integration schemes for Fast Hybrid Testing, *Internal report*, Department of Civil Environmental and Architectural Engineering, University of Colorado at Boulder, CU-NEES-06-8, 5 pages.

CHAPTER 6

Variations in physical test equipment and computer simulation tools for civil structural hybrid simulation

S.M. Jiran
MTS Systems Corporation, Minneapolis, United States

ABSTRACT: For the two primary types of hybrid simulation in civil-structural and seismic engineering, real-time and quasi-static, there is a large variation in the requirements for physical test equipment and computer simulation tools. The variations in the physical test equipment are most evident in the size and capacity of the hydraulic actuation system. The differences in computer simulation relate to the speed for which the models are solved and information is communicated to the physical test system. This paper will review the conventional techniques currently employed to physically test full civil structures; describe how hybrid simulation combines computer simulation with these physical techniques to effectively characterize the properties of sub-structures; and finally explore the basic system architectures of the two types of hybrid simulation that have been successfully deployed by leading universities with the help of MTS Systems Corporation.

1 INTRODUCTION

1.1 *Hybrid simulation technology*

By simultaneously combining physical testing of substructures with computer models of the remainder of the structure, hybrid simulation technology provides a complete picture of how events such as earthquakes can affect large structures such as buildings and bridges without having to physically test the entire structure.

Hybrid simulation technology brings a new level of reality to civil engineering laboratory tests. By integrating computer simulation and physical testing, hybrid simulation more accurately captures the effects that a substructure has on the overall structure, while subjecting the substructure to the forces and motions it would experience if it were in-place within the complete structure.

1.2 *Laboratory test equipment technology*

Most civil-structural and seismic test laboratories have the equipment needed to perform tests ranging from basic material characterizations to full structural evaluations. A typical laboratory tool-set includes hydraulic actuators, control systems, hydraulic pumps, load cells, displacement transducers, connections (swivels or tables) and reaction structures.

Because of cost constraints, most laboratories have equipment designed to perform slower, or quasi-static, tests. The equipment needed to perform dynamic tests, however, requires a large amount of hydraulic power—larger pumps supply large amounts of fluid to actuators with over-sized servovalves. Additionally, these tests require an array of dynamic load and displacement measurement devices and advanced controllers capable of running more sophisticated techniques to effectively control and monitor multiple states of motion.

Figure 1.   Typical test at the national center for research on earthquake engineering in taiwan.

1.3   *Computer simulation tools technology*

Finite element analysis is the primary computer simulation for civil-structural engineering. The finite elements are highly complex to account for the composite materials that are used in civil construction. The analysis techniques are primarily time-based for events like earthquakes and incorporate unique integration algorithms.

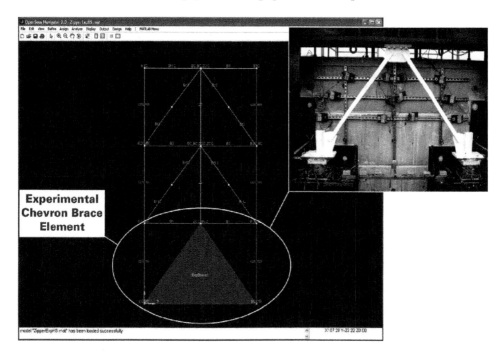

Figure 2.   Depiction of the experimental element concept (Schellenberg).

With the implementation of hybrid simulation the concept of an experimental element has been introduced. The experimental element is the representation of substructure that is to be tested within the computer simulation (see figure below). The experimental element contains the interface to communicate with a physical test (Schellenberg).

2   CONVENTIONAL CIVIL-STRUCTURAL AND SEISMIC TESTING TECHNIQUES

2.1   *Full dynamic testing*

Full dynamic testing is performed by subjecting a civil-structural specimen to the simulated motions of an earthquake event using a seismic simulator, or shake table (see the three-story building diagram below). These tests are conducted in real-time so that the true dynamics of the earthquake are imparted to the structure. The primary challenge of such a test lies in achieving the degree of system control necessary to ensure accurate reproduction of the exact motions of the ground under the test specimen.

The characteristics of the motions generated by the shake table are unique to civil structural and seismic testing as shown in the above figure. The peak displacements are much greater than the vibrations induced in an aerospace or ground vehicle environment.

2.2   *Pseudodynamic testing*

The pseudodynamic test method combines well-established structural dynamics analytical techniques with experimental testing. It is a reliable, economic and efficient method for evaluating large-scale structures that are too large or massive to be tested with a shake table. The base of the structure is fixed in the laboratory and the motions of the earthquake are applied via hydraulic actuators at significant locations throughout the structure.

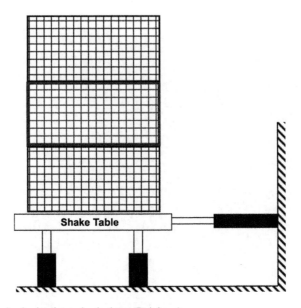

Figure 3.   Typical seismic simulator physical test (Reinhorn).

Earthquake	Peak Ground Acceleration (PGA)	Peak Ground Velocity (PGV)	Peak Ground Displacement (PGD)
Chi-Chi, Taiwan 20-Sep-1999	0.364 g	0.554 m/s	256 mm
Kobe, Japan 16-Jan-1995	0.821 g	0.813 m/s	177 mm
Northridge, CA 17-Jan-1994	0.617 g	0.408 m/s	85.7 mm

Figure 4.   Typical earthquake ground motion characteristics (PEER).

In pseudodynamic testing the complete test structure is first idealized as a discrete-parameter system, so that the motion of the system can be described by second-order ordinary differential equations.

Secondly, assumptions are made for the inertial and viscous damping characteristics of the system according to the figure below. The structural restoring forces, on the other hand, are directly measured during an initial test where force is applied to the structure at a slow, or quasi-static, rate. A step-by-step numerical integration method is used to gather and append these experimental values to the system equations.

Once the equations have been determined by the first experiment, a simulated earthquake excitation is imposed on the test structure with results from the model. Thus, the quasi-statically imposed displacements of the test structure will resemble those that would be generated if the structure were tested dynamically.

The following diagram shows the differences between the two conventional testing methods.

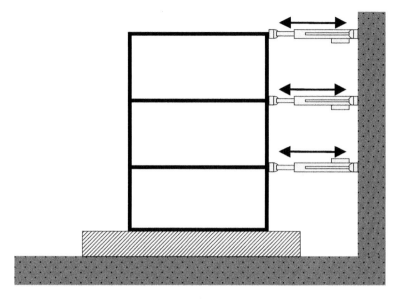

Figure 5.   Typical pseudodynamic physical test (Reinhorn).

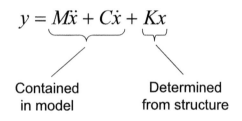

Figure 6.   Equation of motion separation for pseudodynamic technique.

3   CIVIL HYBRID SIMULATION

3.1   *Hybrid simulation*

Hybrid simulation, which can be used to perform both full dynamic and pseudodynamic testing, is defined by the concept of substructuring. The full dynamic and pseudodynamic test methods described above involve the testing of complete structures. The hybrid simulation method includes a physical test of a portion of the structure and the rest of the structure is modeled in a computer simulation. The computer simulation is generally performed with finite elements that will most accurately represent the dynamic properties of the simulated structure.

The physical test recreates the boundary conditions within the computer model. In the above example, the shake table represents the interface between the first and second floor of the building and the structural actuator represents the interface between the second and third floors of the building.

Hybrid simulation can also be broken into the categories of dynamic and pseudodynamic. The above is an example of a full dynamic hybrid simulation with all of the mass being included in the test article. A variation of this simulation can also be conducted with the removal of the non-structural components, or mass components of the physical test specimen. This simulation is called

Full Structure

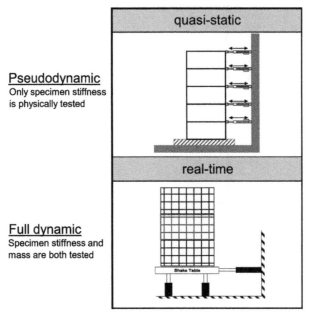

Figure 7.   Comparison of conventional testing techniques.

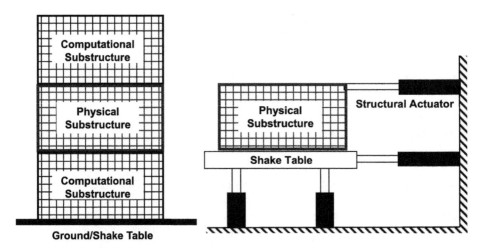

Figure 8.   Hybrid simulation example (Reinhorn).

real-time pseudodynamic where the damping properties of the test specimen remain apart of the physical specimen according to the following equation.

The full spectrum of civil structural techniques is shown in the following chart. This chart illustrates the differences between conventional and hybrid simulation (sub-structuring) and the contrast between pseudodynamic and dynamic simulation as well.

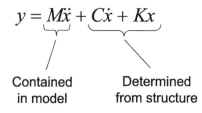

$$y = \underbrace{M\ddot{x}}_{} + \underbrace{C\dot{x} + Kx}_{}$$

| Contained | Determined |
| in model | from structure |

Figure 9.   Equation of motion separation for real-time pseudodynamic technique.

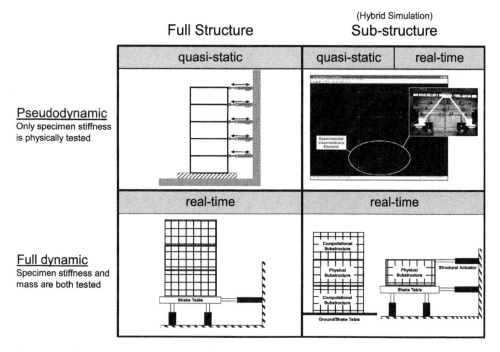

Figure 10.   Comparison of conventional and hybrid simulation testing techniques.

4   QUASI-STATIC HYBRID SIMULATION

4.1   *Motivation*

Quasi-static hybrid simulation is used to evaluate substructures that predominantly contribute stiffness and strength to the complete structure. The forces and motions in a quasi-static simulation are applied at an artificially slow rate to allow for a more detailed study of a structure and to accommodate the limited capacity of the hydraulic actuators and pumping systems of most civil engineering laboratories.

The typical system set-up for a quasi-static hybrid simulation includes a computer simulation PC which runs the finite element model. Commands for the physical test are communicated to the test PC via an ethernet (TCP/IP) connection. The test PC interprets the commands and sends a signal to the servocontroller which in-turn activates a closed-loop control algorithm to move the actuator to the desired location. The resulting state from the computer simulation command is then communicated back-up the same chain to the computer simulation PC.

4.2   *Physical test equipment*

The physical test equipment required for a quasi-static hybrid simulation is more typically found in a civil structural laboratory and is also used to perform more conventional tests. Hydraulic actuators are designed for slow cycles (maximum 1–2 Hz), and the requirement of hydraulic fluid flow to each of the actuators is typically in the range of 115 to 230 liters per minute. The operation of three or four of these actuators simultaneously in a hybrid simulation is achievable with the pumping systems found in most civil structural laboratories.

Displacement and force measurement transducers must also measure relatively slow changes in position and load. A magnetostrictive linear-position transducer is typically used for displacement measurement, and a load cell is typically used for force measurement.

4.3   *Computer simulation tools*

There is a wide choice of computer simulation tools based in finite elements that are typical used for the analysis of civil structures. The implementation of hybrid simulation is dependent on the ability to create a new experimental element in the finite element code. The speed of a hybrid simulation is highly dependent on the speed of which the finite element code can perform the calculations to change the state of the model from current to future time steps.

To bridge between the computer simulation and physical test environments, a communication framework is needed to exchange command and feedback signals. Most typically the signals are

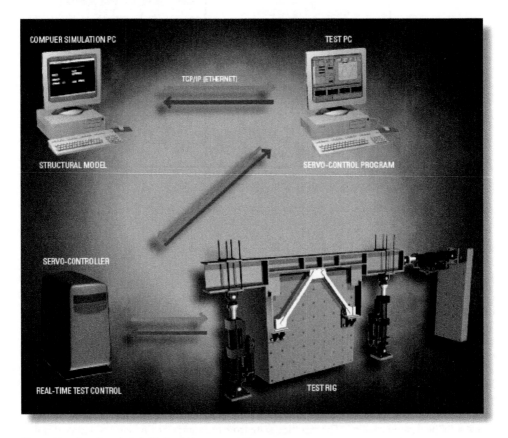

Figure 11.   Quasi-static hybrid simulation system diagram.

sent between computers via a TCP/IP connection which can be within the same laboratory or geographically distributed. Another important function of the framework is to map the control points and degrees of freedom in the computer simulation to the actuator control channels in the physical test equipment.

5 REAL-TIME HYBRID SIMULATION

5.1 *Scope of the simulation*

Real-time hybrid simulation is used to evaluate substructure or components that contribute damping or inertia effects to the complete structure. A real-time hybrid simulation requires high-force, dynamic structural actuators and large hydraulic power units to evaluate all dynamic properties (mass, damping & stiffness) of the substructure. Real-time hybrid simulation also means that the computer simulation is performed in real-time. The computer model is developed in a tool such as Simulink® and downloaded to a real-time target PC. A special high-speed connection is made directly to the real-time test controller via SCRAMNet® reflective memory.

The difficulty of accurately controlling the physical test escalates when performing a real-time hybrid simulation. The test controller must react instantly to the computer simulation while maintaining three parameters of motion (displacement, velocity & acceleration).

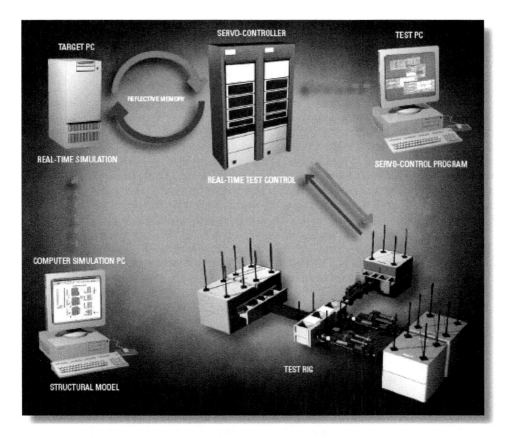

Figure 12.   Real-time hybrid simulation system diagram.

5.2   *Physical test equipment*

Real-time requires high-speed, high-force hydraulic actuators along with a large amount of hydraulic pump capacity to deliver appropriate forces and motions. A real-time hybrid simulation requires high-force, dynamic structural actuators and large hydraulic power units to evaluate all dynamic properties (mass, damping & stiffness) of the substructure. Actuation is achieved with actuators designed for high-frequency, high-speed fatigue tests. Displacement and load measurements are provided by LVDTs and delta-pressure cells, respectively.

5.3   *Computer simulation tools*

The system's Computer Simulation PC features real-time dynamic modeling software, such as Simulink, which can run in real-time when downloaded to the system's xPC Target® PC. Simulink is a general purpose dataflow programming environment, widely used in universities and industry worldwide. After developing a Simulink model on the Computer Simulation PC, Real-Time Workshop® can be used to download it to the xPC Target PC where a connection is established to the real-time test controller via reflective memory. The xPC Target PC runs a kernel that provides deterministic performance on PC hardware for running real-time models. High performance is achieved by booting the kernel rather than DOS or Windows. The xPC Target kernel is tuned for minimal overhead and maximum performance with published sample rates approaching 100 kHz.

6   CONCLUSIONS

The two approaches to hybrid simulation require significantly different physical test equipment and computer simulation tools. Quasi-static hybrid simulation leverages physical test equipment and computer simulation tools that are commonly found in civil structural laboratories. Real-time hybrid simulation requires specialized tools for both physical teat equipment and computer simulation that are focused on providing maximum speed.

Most civil engineering research institutions will benefit from degree of hybrid simulation technology. By integrating computer analysis and physical testing, hybrid simulation enables them to efficiently model a significant portion of a structure on a computer, while simultaneously conducting physical tests on critical substructures or components, thus eliminating the need to perform expensive large-scale physical tests on full structures using multi-degree-of-freedom seismic simulators. Only a handful of labs, however, have the resources to set up and maintain effective hybrid simulation facilities on their own. To realize the full benefits of hybrid simulation, the rest require more user-friendly solutions that minimize IT requirements and maximize their lab's research potential by facilitating the sharing of test data across projects and peer institutions.

Through the collaboration of leading universities in the NEES Program—including the University of California Berkeley, State University of New York Buffalo, the University of Colorado Boulder, the University of Nevada-Reno—and industry partners such as MTS, an array of affordable, high-performance hybrid simulation solutions is available to meet the technical and budgetary requirements of a wide range of civil engineering labs.

REFERENCES

Schellenberg, A., and Mahin, S., 2006. Integration of Hybrid Simulation within the General-purpose computational Framework OpenSees, Proceedings, 8th U.S. National Conference on Earthquake Engineering, San Francisco.

Reinhorn, A.M., Sivaselvan, M.V., Liang, Z., and Shao, X., 2004. Real-time Dynamic Hybrid Testing of Structural Systems, Proceedings, 13th World Conference on Earthquake Engineering, Vancouver.

Pacific Earthquake Engineering Research Center Strong Motion Database http://peer.berkeley.edu/smcat/search.html

CHAPTER 7

Real-time hybrid experimental system with actuator delay compensation

Y. Dozono
Hitachi, Ltd., Ibaraki, Japan

T. Horiuchi
Hitachi Plant Technologies, Ltd., Tokyo, Japan

ABSTRACT: An actuator-delay compensation method was developed for a real-time hybrid experimental system. The response delay of an actuator is equal to a negative damping, and therefore, causes an experimental error. This method predicts the displacement of the actuator after a response delay time using the extrapolation of previous displacements, and corrects the command signal to the actuator. Also presented in this paper is the stable criterion of this method, which depends on the response delay time and a natural frequency of the structural system under consideration. The validity of this criterion was confirmed and the effectiveness of the method was verified through a series of real-time hybrid experiments for a four-degree-of-freedom system.

1 INTRODUCTION

It is important to design a structure considering deformation performance extended to a failure under dynamic loading such as an earthquake because of simultaneous pursuit of economic efficiency and safety. Therefore, it is necessary to clarify the response characteristics of structures, such as a nonlinear load-deformation curve and velocity dependence.

Numerical analysis is one of evaluation techniques for a structural design. A numerical analysis is suitable when the characteristics of all the elements under actual load and boundary conditions are accurately known. However, this precondition is not satisfied, for example, if new devices or materials are used or an element is used under new load and/or boundary conditions. Thus, a shaking table test is often conducted to evaluate the seismic response and build a model of the element. In a shaking table test, the whole structure is excited at a desired acceleration, but the size of the specimen (the excited structure) is limited due to the load capacity of the shaking table. Thus, a real-time hybrid experimental method has been developed (Hakuno et al., 1969, Nakashima et al., 1992, Horiuchi et al., 1993).

In a real-time hybrid experiment, the structure to be evaluated is separated into two parts; a numerical model and a specimen. The response of the numerical model is calculated based on the tested results of the specimen, and the specimen is excited based on the analyzed results of the numerical model. In other words, the interactions between the numerical model and the specimen reflect on each other.

For an accurate real-time hybrid experiment, the boundary conditions of the numerical model must agree with those of the specimen in the spatial and temporal domains. A spatial agreement is achieved by the design of the supports of the specimen and the arrangement of the actuators. A temporal disagreement is caused by the response delay of an actuator and the data transfer time between a computer and an actuating system. The data transfer time will be short enough and negligible when using memory sharing devices, such as Supper Real-Time Controller (Umekita et al., 1995). Therefore, we developed a compensation method for the response delay of an actuator (Horiuchi et al., 1995, 1996a, 1996b, 1999, 2001). We introduce this compensation method in this paper and show its effectiveness through a series of real-time hybrid experiments for a four-degree-of-freedom structure.

2   OVERVIEW OF HYBRID EXPERIMENT

In a real-time hybrid experiment, the dynamic loading test of a specimen and a numerical analysis are simultaneously executed and linked to each other. Real-time hybrid experimental systems can be classified into two groups. In the first type, a calculation is made to find the force that should be applied to a specimen by an actuator (Dimig et al., 1999, Reinhorn et al., 2004). After the specimen is deformed, the displacement and velocity of the specimen are measured and input to the next-step calculation. In the second type, a calculation is made to find the displacement of a boundary between the numerical model and the specimen. After the specimen is deformed, a reaction force from the specimen is measured and is input to the next-step calculation. The second type is suitable for tests on mechanical devices, such as dampers and seismic isolators, because we can choose only the element whose characteristics are difficult to express with a numerical model as a specimen. We will explain the compensation method applied to the second type of real-time hybrid experiment in this paper.

A conceptual view of a real-time hybrid experiment is shown in Fig. 1. The original structure, of which the seismic response is of interest, is divided into two parts. One is a specimen (a physical model), which is excited with an actuator, and the other is a numerical model. The behavior of the numerical model is calculated based on the following equation of motion:

$$M_n\ddot{x} + C_n\dot{x} + K_nx = f + q, \tag{1}$$

where M_n, C_n, and K_n are the mass, damping, and stiffness matrices, respectively, of the numerical model. x is a relative displacement vector, f is an external force vector, such as an earthquake load, and q is the reaction force vector generated by the specimen. The dot represents differentiation with respect to time. The reaction force vector, q, is formally written using the displacement at the boundary, x_b, as follows:

$$q = q(x_b, \dot{x}_b, \ddot{x}_b, \cdots). \tag{2}$$

Therefore, by repeating the following steps, the seismic response of the whole structure can be evaluated: (a) measure the reaction force, q, from the dynamic loading test; (b) calculate the vibration response of the numerical model using the measured reaction force vector, q, and the predetermined external force vector, f, and (c) excite the specimen based on the calculated vibration response, x. From Eq. (2), it can be understood that it is necessary to conduct the dynamic loading

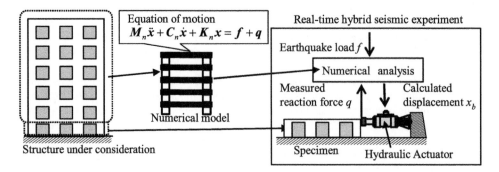

Figure 1.   Conceptual view of real-time hybrid experiment.

test and the computer calculation on a common time axis (that is, in real time) if the reaction force largely depends on the displacement derivatives with respect to time.

3   ACTUATOR-DELAY COMPENSATION

3.1   *Dynamic characteristics of hydraulic actuators*

Hydraulic actuators are preferable for applying the hybrid experimental method to a large-scale structure, because a large actuator-excitation force is required. However, the dynamic characteristics of hydraulic actuators have a small but inevitable response delay. A representative example of the response delay characteristics is shown in Fig. 2, which compares a commanded displacement signal to the resulting displacement.

The response delay is equivalent to negative damping in a real-time hybrid experiment. This is explained as follows. Let us take into consideration a single-degree-of-freedom (SDOF) system shown in Fig. 3(a), where a spring with stiffness of k, is under excitation as shown in Fig. 3(b). Let ω_0 and δt be the natural frequency of the SDOF system and the response delay time of the actuator, respectively, and assume that the damping is zero for simplicity. When the system is in free vibration with a displacement amplitude, A, the calculated displacement of the mass can be written as follows:

$$x = A \sin \omega_0 t. \tag{3}$$

And the actual excited displacement, x'' becomes

$$x'' = A \sin(\omega_0 t - \omega_0 \delta t). \tag{4}$$

The reaction force, q, is proportional to the actual excited displacement, x'', not to the calculated displacement of the mass, x. Then the reaction force can be written as follows:

$$q = kx''. \tag{5}$$

Therefore, by assuming that δt is small, the change in the total system energy per period, δE, becomes

$$\delta E = \oint q dx = \int_0^T q \frac{dx}{dt} dt = \int_0^T (kx'') \frac{dx}{dt} dt$$
$$= \int_0^T kA \sin(\omega_0 t - \omega_0 \delta t) \cdot A \cos \omega_0 t \, dt = \frac{1}{2} kA^2 \cdot 2\pi \omega_0 \delta t \tag{6}$$

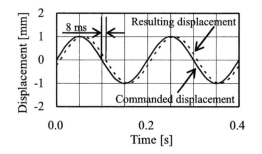

Figure 2.   Time history of commanded and resulting signals for a hydraulic actuator.

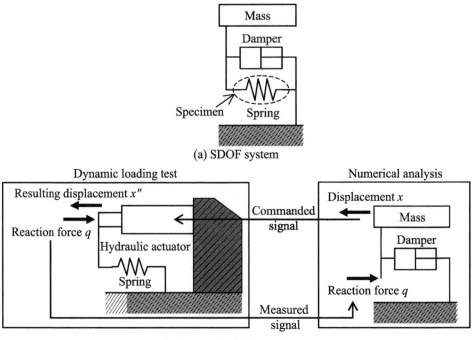

(a) SDOF system

(b) Setup of hybrid experiment

Figure 3.   Real-time hybrid experimental system for a single-degree-of-freedom system.

where $T = 2\pi/\omega_0$. This means that the response delay of the actuator in a real-time hybrid experiment causes an increase in the total energy. This energy increase is the same as that caused by equivalent negative damping, c_{eq}; that is,

$$c_{eq} = -k\delta t. \tag{7}$$

The negative damping causes an error in a hybrid experiment. Moreover, the response (the numerical analysis) will diverge and the experiment will become impossible to complete, if the negative damping is larger than the inherent structural damping.

3.2   Method for compensating actuator delay

The developed compensation method predicts the displacement of an actuator after the response delay time, δt. In order to shorten the prediction-calculation time to accomplish a real-time excitation, the displacement is predicted with the following simple equation:

$$x' = \sum_{i=0}^{n} a_i x_i, \tag{8}$$

where x' is the predicted displacement, n is the order of the prediction, x_0 is the present calculated displacement, x_i is the calculated displacement at the $\delta t \times i$ unit of time ago, and a_i are the constants listed in Table 1. In the equation, the predicted value, x', is obtained by extrapolating the nth-order polynomial function based on the present and n previous calculated values, as shown in Fig. 4. As an example, the extrapolation function for $n = 1$ is a first-order (linear) function. Let $P(t)$ and

Table 1.   Constants in prediction equation.

Order n	a_0	a_1	a_2	a_3	a_4
0	1	–	–	–	–
1	2	−1	–	–	–
2	3	−3	1	–	–
3	4	−6	4	−1	–
4	5	−10	10	−5	1

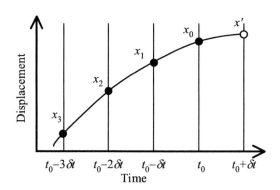

Figure 4.   Schematic of compensation method.

t_0 be the displacement at time, t, and the present time, respectively. A first-order extrapolation function is written as Eq. (9).

$$\frac{P(t) - P(t_0)}{t - t_0} = \frac{P(t_0) - P(t_0 - \delta t)}{\delta t} \tag{9}$$

The predicted displacement at $t = t_0 + \delta t$, x', can be calculated by substituting $t = t_0 + \delta t$ into Eq. (9) such that

$$x' = P(t_0 + \delta t) = 2P(t_0) - P(t_0 - \delta t) = 2x_0 - x_1 \tag{10}$$

This formula becomes identical to the one in Eq. (8), and thus constants, a_0 and a_1, are equal to those listed in Table 1. The constants, a_i, for the other values of n are obtained in a similar manner.

3.3   *Limitation of proposed method*

Since a perfect prediction is impossible, some limitation must exist in the application of such compensation. The limitation will be discussed by considering an experiment on a single-degree-of-freedom model. By introducing the prediction subsystem on the SDOF experiment shown in Fig. 3, the control signal to the actuator becomes the predicted value, x', instead of the calculated value, x. Let the calculated vibration response, x, be a sinusoidal wave with a circular frequency of ω, and an amplitude of A, i. e. $x = A \sin \omega t$. The actual-excited displacement, x'', obtained through the prediction subsystem written in Eq. (8), and the actuator with a response delay time of δt becomes

$$x'' = \sum_{i=0}^{n} a_i \sin\{\omega t - (i + 1)\omega \delta t\}. \tag{11}$$

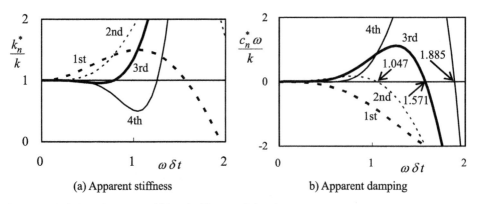

Figure 5.   Variation of apparent additional stiffness and damping.

By considering the fact that $\dot{x} = A\omega \cos \omega t$, the reaction force used in calculation, q, can be written as follows:

$$q = kx'' = k_n^* x + c_n^* \dot{x}, \tag{12}$$

where k_n^* and c_n^* are the apparent stiffness and damping, respectively, for the nth-order prediction, they are written as follows:

$$\frac{k_n^*}{k} = \sum_{i=0}^{n} a_i \cos(i+1)\omega\delta t, \quad \text{and} \tag{13}$$

$$\frac{c_n^*\omega}{k} = -\sum_{i=0}^{n} a_i \sin(i+1)\omega\delta t. \tag{14}$$

From Eqs. (13) and (14), it is obvious that the apparent stiffness, k_n^*, and damping, c_n^*, are functions of $\omega\delta t$, as shown in Fig. 5. The value k_n^*/k is almost equal to one and the value $c_n^*\omega/k$ is almost equal to zero. Therefore, the prediction is almost ideal when $\omega\delta t$ is small, although the prediction causes small variations in both the stiffness and damping. It should be noted, however, that the damping becomes negative, and thus the calculation diverges, when the natural frequency of the whole structure is high or the response delay time is long, so that the value of $\omega\delta t$ is larger than a critical value which depends on the prediction order, n. This is the limitation of the developed compensation method. In the experiments discussed below, we will use the third-order prediction, which is drawn with solid lines in Fig. 5, because it requires only a small computational load and produces a large critical value of $\omega\delta t$ (1.571) for a stable calculation.

As discussed above, the stability in the calculation depends on the non-dimensional value, $\omega\delta t$. Here, the response delay time, δt, is affected by an exciting frequency, an exciting amplitude, and the vibration characteristics of a specimen. However, in this paper, we treated δt as a constant because the change in δt is sufficiently small when compared with the value of δt.

3.4   *Application to a multi-degree-of-freedom system*

In this section, we discuss the stability criterion for a multi-degree-of-freedom (MDOF) system. Consider the structure shown in Fig. 6, in which a numerical model is a MDOF system. The equation

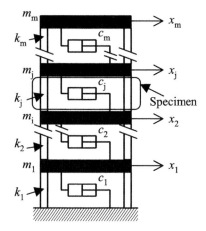

Figure 6.　Multi-degree-of-freedom system.

of motion for the structure is written as follows:

$$M\ddot{x} + C\dot{x} + Kx = f, \tag{15}$$

where M, C, and K are the mass, damping, and stiffness matrices, respectively, of the whole structure. x is a relative displacement vector and f is an external force vector. The specimen is a spring element between mass points, i and j. In addition, its stiffness is k_j. Let us introduce here a matrix, B, determined by the position of the specimen in the whole structure. The elements of B are

$$B_{mn} = \begin{cases} 1 & (m = n = i \quad or \quad m = n = j) \\ -1 & (m = i, n = j \quad or \quad m = j, n = i). \\ 0 & (otherwise) \end{cases} \tag{16}$$

Then, Eq. (15) can be written as follows:

$$M\ddot{x} + C\dot{x} + (K - k_j B)x + k_j Bx = f. \tag{17}$$

Let x_{ij} be the relative displacement of mass point i to mass point j. The fourth term on the left hand side can be written as

$$k_j Bx = ik_j x_{ij}, \tag{18}$$

where i is a vector whose ith element is 1, jth element is -1, and the other elements are 0.

Here, x_{ij} is a calculated displacement and is not the same as the actual-excited displacement, x_{ij}''. Therefore, Eq. (18) is replaced by Eq. (19),

$$ik_j x_{ij}'' = i(k_j^* x_{ij} + c_j^* \dot{x}_{ij}) = B(k_j^* x + c_j^* \dot{x}), \tag{19}$$

where k_j^* and c_j^* are functions of $\omega \delta t$. By substituting Eq. (19) to Eq.(17), we can obtain Eq. (20).

$$M\ddot{x} + (C + c_j^* B)\dot{x} + \{K - (k_j - k_j^*)B\}x = f \tag{20}$$

Based on Eq. (20), let us consider the variations in apparent stiffness and apparent damping of the specimen concerning the rth-order eigenmode. If we assume that k_j^* and c_j^* are small, then the variation in the eigenmode vector, ϕ_r, is small. The apparent circular frequency, ω'_r, and the apparent additional damping, ς'_r, of the specimen can be approximately written as:

$$\omega'^2_r = \frac{\{\phi_r^T \boldsymbol{K} \phi_r - k(1 - k_j^*/k_j)\phi_r^T \boldsymbol{B} \phi_r\}}{\phi_r^T \boldsymbol{M} \phi_r} = \omega_r^2 \left\{1 - p_r\left(1 - \frac{k_j^*}{k_j}\right)\right\}, \quad \text{and} \tag{21}$$

$$\varsigma'_r = (c^* \omega'_r / 2k_j) p_r (\omega_r / \omega'_r)^2, \tag{22}$$

where $\omega_r^2 = \phi_r^T \boldsymbol{K} \phi_r / \phi_r^T \boldsymbol{M} \phi_r$ and $p_r = (k_j/\omega_r^2)(\phi_r^T \boldsymbol{B} \phi_r / \phi_r^T \boldsymbol{M} \phi_r)$ are constants. The parameter p_r represents the degree of the influence affected by the specimen to the rth-order eigenmode. Although p_r is equal to one for a SDOF system, it is usually less than one for a MDOF system. Thus, the variation in the apparent stiffness and the apparent additional damping of the specimen caused by the actuator response delay is most significant on a SDOF system.

The stability limit of a real-time hybrid experiment for a MDOF system is obtained as follows: (a) Identify the stiffness of a specimen using a static loading test. (b) Calculate ω_r and p_r using an eigenvalue analysis of the whole structure. (c) Calculate $\omega'_r \delta t$ to satisfy $\varsigma'_r = 0$, which is the stability limit, using Eq. (22). (d) Substituting k_j^* at this condition to Eq. (21), calculate ω'_r at the stability limit. (e) Calculate the limit of the response delay time, δt_{cr}, by dividing $\omega'_r \delta t$ by ω'_r. Here, it is considered that the stability limit for the eigenmode where p_r is sufficiently small can be neglected because the negative additional damping of the specimen will be cancelled by the damping of the whole structure.

In the above-mentioned procedure, the stability of a real-time hybrid experiment can be evaluated using Eq. (15) and the response delay time of an actuator.

4   VERIFICATION TEST OF COMPENSATION METHOD

4.1   *Experimental method*

We conducted a series of experiments to evaluate the stability limitation and effectiveness of the compensation method. Figure 7 shows a target structure that consists of four mass points and four units of springs. Although the structure was a six-degree-of-freedom system consisting of the translations of four mass points and the rotations of two mass points, the rotational freedom was condensed, and four degrees of freedom were intended (Guyan, 1965). Figure 8 shows the condensed numerical model. In the experiments, the spring element between the basement and the mass point 1 was replaced with a specimen, and the other elements were replaced with a numerical model. Table 2 shows results of the eigenvalue analysis for the condensed numerical model. Figure 9 shows a schematic of the experiment. In the vibration analysis block, Eq. (1) for the numerical model is solved, and the displacement of mass point 1, x, is calculated. x is transformed to the predicted displacement, x', in the delay compensation block. The hydraulic actuator is excited to acomplish x'. However, the actual-excited displacement is not x' but x'' because of the response delay. To clearly evaluate the proposed method in the experiments, the hydraulic actuator did not excite a specimen. The product of x'' and the stiffness of the specimen was fedback to the vibration analysis instead of a reaction force. Here, the time interval of the vibration analysis and the delay prediction was 0.5 ms.

4.2   *Evaluation of stability limit*

To evaluate the stability limitation, the hydraulic actuator was replaced with an electrical delay element whose delay time was adjustable and free vibration tests are conducted. The same initial velocity was generated to all the mass points, and all the eigenmodes were excited. The results of the free vibration tests for different delay times are as follows; the displacement of mass point 1

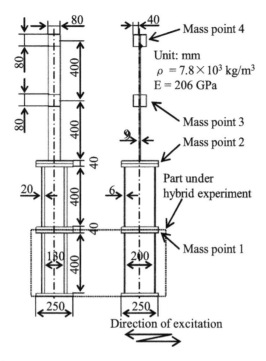

Figure 7. Experimental structure.

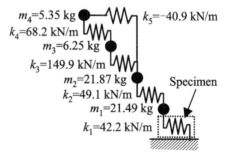

Figure 8. Numerical model of structure.

Table 2. Results of eigenvalue analysis of structure.

Mode no.	1	2	3	4
Natural freq. [Hz]	3.49	6.16	12.03	33.34
Parameter p_r	0.470	0.243	0.239	1.70×10^{-5}

was not diverged when $\delta t \leq 12.5$ ms, and was diverged when $\delta t = 13.5$ ms, as shown in Fig. 10. Based on the procedure presented in the previous section, the limit of the response delay time for the maximum-order (4th order) eigenmode was 7.52 ms, as shown in Table 3. This is much smaller than the value from the experimental results. It is considered that the specimen only slightly deforms at the fourth-order eigenmode and the limitation depends on the third-order eigenmode,

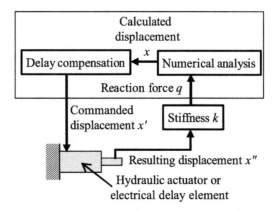

Figure 9.   Schematic of experiment.

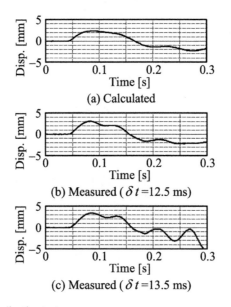

Figure 10.   Results of free vibration test.

because of a small parameter, p_r, at the fourth-order eigenmode, as shown in Table 2. The limit value for the third-order eigenmode shown in Table 3 agrees with the experimental results. From these experiments, it is verified that the evaluation of the stability limitation is appropriate.

4.3   *Evaluation of compensation method*

To evaluate how effective the compensation method was, the results form the hybrid experiments using a hydraulic actuator were compared with those using the electrical delay element whose delay time was zero (that is, $x = x''$). The acceleration of the basement was sinusoidal waves of ten periods. Figure 11 shows an example of time history waves of displacements at mass points, 1 and 3. Since the compensation works well, the apparent delay time of the hydraulic actuator becomes almost zero, and then both results agree with each other. Figure 12 shows the maximum displacements of mass point 3 divided by the maximum accelerations of the basement. Both results are in excellent agreement with each other over the tested frequency range.

Table 3.   Critical delay time for structure.

Mode no.	3	4
Critical delay time [ms]	12.44	7.52

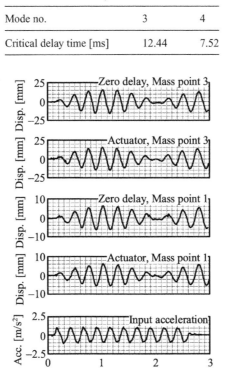

Figure 11.   Displacement time history for ten-period harmonic excitation.

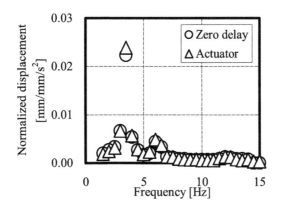

Figure 12.   Maximum displacement for ten-period harmonic excitation.

From these results, we proved that the propose method can compensate for the response delay of a hydraulic actuator, therefore enables a highly accurate real-time hybrid experiment.

5   CONCLUSION

In a hybrid experiment where the dynamic loading test of a specimen and a numerical analysis of the remaining parts are linked, the response delay of an actuator decreases the accuracy of the test. Therefore, we developed a compensation method for the response delay, and evaluated its effectiveness through a series of real-time hybrid experiments.

The obtained results can be summarized as follows:

- The method has a stable criterion that depends on a natural frequency of the structural system under consideration and the delay time of an actuator. The validity of the criterion was confirmed through a series of experiments.
- The results of real-time hybrid experiments using the developed compensation with those of a numerical analysis were compared. The results had good correlation, and therefore, the effectiveness of this method was verified.

Moreover, we have been improving the compensation method to expand the coverage of a real-time hybrid experimental method. The allowable ratio of the mass of the specimen to that of the numerical model increases about three times compared with the presented method (Horiuchi et al., 2001).

REFERENCES

Dimig, J., Shield, C., French, C., Bailey, F. & Clark, A., 1999, Effective Force Testing: A Method of Seismic Simulation for Structural Testing, J. of Struct. Engng., Vol. 125, No. 9, 1028–1037.

Guyan, R. J., 1965, Reduction of Stiffness and Mass Matrics, AIAA J., Vol. 3, No. 3, 380

Hakuno, M., Shidawara, M. & Hara, T., 1969, Dynamic Destructive Test of a Cantilever Beam Controlled by an Analog Computer, Trans. Japan Society of Civ. Engrs., Vol. 171, 1–9 (in Japanese)

Horiuchi, T., Nakagawa, M. & Kametani, M., 1993, Development of Real-Time On-Line Vibration Testing System for Seismic Experiments, Trans. SMiRT-12, 337–342

Horiuchi, T., Nakagawa, M., Sugano, M. & Konno. T., 1995, Development of a Real-Time Hybrid Experimental System with Actuator Delay Compensation (1st Report), Trans. Japan Society of Mech. Engrs., Vol. 61, No. 584, 1328–1336 (in Japanese)

Horiuchi, T., Nakagawa, M., Sugano, M. & Konno. T., 1996a, Development of a Real-Time Hybrid Experimental System with Actuator Delay Compensation, Proc. 11th World Conf. on Earthquake Engineering, Paper No. 660

Horiuchi, T., Nakagawa, M., Sugano, M. & Konno. T., 1996b, Development of a Real-Time Hybrid Experimental System with Actuator Delay Compensation (2nd Report), Trans. Japan Society of Mech. Engrs., Vol. 62, No. 599, 2563–2570 (in Japanese)

Horiuchi, T., Inoue, M., Konno. T. & Namita, Y., 1999, Real-Time Hybrid Experimental System with Actuator Delay Compensation and its Application to a Piping System with Energy Absorber, Earthquake Engng. Struct. Dyn., Vol. 28, 1121–1141

Horiuchi, T. & Konno. T., 2000, Development of a Real-Time Hybrid Experimental System with Actuator Delay Compensation (4th Report), Trans. Japan Society of Mech. Engrs., Vol. 66, No. 650, 3225–3232 (in Japanese)

Horiuchi, T. & Konno. T., 2001, Development of a Real-Time Hybrid Experimental System with Actuator Delay Compensation (5th Report), Trans. Japan Society of Mech. Engrs., Vol. 67, 667–675 (in Japanese)

Horiuchi, T. & Konno. T., 2001, A New Method for Compensating Actuator Delay in Real-Time Hybrid Experiments, Phil. Trans. R. Soc. A, Vol. 359, 1893–1909

Reinhorn, A. M., Sivaselvan, R. V., Weinreber, S. & Shao, X., 2004, Real-Time Dynamic Hybrid Testing of Structural Systems, 13th World Conf. on Earthquake Engneering, Paper No. 1644

Umekita, K., Kametani, M. & Miyake, N., 1995, Development of Super Real-Time Controller Which Can Perform Analysis Along with Measurement/Control in Real Time, Proc. 5th Robot Symp. Robotics Society of Japan, 55–58 (in Japanese)

Laboratory facilities

CHAPTER 8

Continuous pseudo-dynamic testing at ELSA

P. Pegon, F.J. Molina & G. Magonette
European Laboratory for Structural Assessment, Joint Research Centre, Ispra, Italy

ABSTRACT: The ELSA laboratory is equipped with a large reaction-wall facility and has acquired its best expertise on the development and implementation of innovative experimental techniques mainly related to testing large-scale specimens by means of the pseudo-dynamic method. Relevant achievements within the testing techniques, such as the continuous pseudo-dynamic test, the implementation of monolithic or distributed substructuring and the development of active control systems, have been obtained thanks to an accurate, home-designed, control system. Its role of reference laboratory in Europe has allowed ELSA to benefit from the collaboration of many prominent research institutions within international projects, providing the maximum scientific added value to the results of the tests.

1 INTRODUCTION

The European Laboratory for Structural Assessment (ELSA) has substantially contributed to new developments within the PsD methodology thanks to a proper in-house design of hardware and software in which high accuracy sensors and devices are used under a flexible architecture with a fast intercommunication among the controllers as highlighted in section 3. The loading capabilities of ELSA's reaction wall are shown in Figure 1 (Donea et al., 1996).

The PsD method is an hybrid technique by which the seismic response of large-size specimens can be obtained by means of the on-line combination of experimental restoring forces with analytical inertial and seismic-equivalent forces (Takanashi & Nakashima, 1986). Thanks to the use of

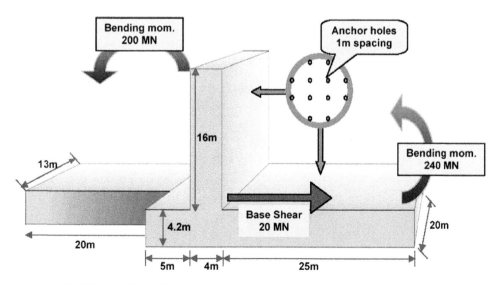

Figure 1. The ELSA reaction wall.

quasistatic imposed displacements, the accuracy of the control and hence the quality of a PsD test is normally better than for a shaking-table test, especially for heavy and tall specimens. In the classical version of the PsD method, displacements are applied stepwise allowing the specimen to stabilise at every step (see section 2). The quality of the test can be further improved using a continuous version of the method as described in section 2. In any case, it is important to recognise that the PsD method is a sophisticated tool that may fail or produce inaccurate results in some cases depending on the systematic experimental errors (Shing & Mahin, 1987, Combescure & Pegon, 1997, Molina et al., 2002a). Particularly, it is well known that the slight phase lags of the control system, which is used to quasi-statically deform the specimen, may considerably distort the apparent damping characteristics of the PsD response and artificially excite the higher modes. The ELSA team has undergone many relevant PsD tests, most of which have been used for the improvement of EuroCode 8 and the assessment of several categories of structures. Taking advantage of this activity, various analysis techniques have been developed and applied that try to assess the magnitude and consequences of the existing experimental errors. Some of these techniques are based on the identification of linear models using a short-time portion of the response. The identified parameters are then transformed into frequency and damping characteristics. By gaining experience in the application of these analysing techniques, it is possible to have a better knowledge of how the feasibility and the accuracy of the experiment depend on the type of structure and the applied PsD testing set-up (Molina & Géradin, 2007).

A different line of application is also worth mentioning, i.e. the use of the PsD method on structures seismically protected by passive isolators or dissipators. In such applications involving new materials and/or devices, the strain rate effect can become significant. Such effect should be reduced there by increasing the testing speed if possible and by an analytical on-line compensation of it when appropriate (De Luca et al., 2001, Molina et al., 2002b, 2004).

Also important to mention are the substructuring techniques developed within the PsD method, which have proved to be very useful for obtaining the seismic response of large structures such as bridges. In that case, a testing set-up is devised in which a limited part of the structure that has the strongest non-linear behaviour (typically some of the piers) is the actual specimen and the rest of the structure (the deck and the remaining piers) is numerically substructured (Pinto et al., 2004). Some details about the current work regarding this topic are given in Section 4. In a different field of research, an important activity of ELSA is also dedicated to real dynamic tests oriented to the development of active-control systems, such as the case of attenuation of vibration on bridges (Magonette et al., 2001, Casciati et al., 2006), vibration monitoring for damage detection or fatigue testing on large cable specimens. The common denominator of all these tests has been the use of innovative testing techniques or the development of advanced structural systems. Finally, during the last years ELSA has also put an important effort in the development of techniques for telepresence, teleoperation and, in general, distributed laboratory environment, which are becoming compatible with the Network for Earthquake Engineering Simulation (NEES) (Pinto et al., 2006). NEES currently integrates the major US laboratories, making it possible for researchers to collaborate remotely on experiments, computational modelling, and education.

2   THE CLASSICAL AND THE CONTINUOUS PSD METHOD

PsD testing consists of the step-by-step integration of the discrete-DoF equation of motion

$$\mathbf{Ma} + \mathbf{r(d)} = \mathbf{f}(t) \qquad (1)$$

Where $\mathbf{M}$ is the theoretical matrix of mass, $\mathbf{a}$ and $\mathbf{d}$ represent the unknown vectors of acceleration and displacement and $\mathbf{f}(t)$ are known external forces that, in the case of a seismic excitation, are obtained by multiplying the specified ground acceleration by the theoretical masses. The unknown restoring forces $\mathbf{r(d)}$ are experimentally obtained at every time integration step by quasistatically imposing, generally by means of a hydraulic control system, the computed displacements. Note that Equation 1 does not involve a damping matrix. This is to underline that, in most of the cases, the damping matrix is useless at the laboratory since it is just a straightforward artefact which is used in pure analytical simulation to introduce an equivalent model for micro-hysteresis dissipation. For every PsD test, we will define the prototype or accelerogram time t (Figure 2), which corresponds to the one of the original problem with an earthquake excitation that may last for a few tens of seconds, and the experimental time T, which may extend to several hours for the execution of the test in the laboratory. We may call λ the time scale factor, defined as the proportion between experimental and prototype times, i.e $\lambda = T/t$.

In the classical PsD method, the time increment Δt for the integration of the equation of motion is chosen small enough to satisfy the stability and accuracy criteria of the integration scheme. The ground accelerogram must be discretized with the prototype record increment (Figure 2). The smaller this time increment is chosen, the larger the number of integration steps will be to cover the duration of the earthquake. The execution of every step will take place in the corresponding experimental time lapse, which uses to be in the order of several seconds. In fact, the experimental time ΔT is split in four phases (Figure 2 left):

- A stabilising hold period ΔT_{h1} of the system motion after the ramp of the reference signal at the controller. In practise, this period allows the specimen to reach the computed displacement. If the computed displacement has not been accurately achieved, the measured force will not correspond to the computed displacement.
- A measuring hold period ΔT_{h2} that allows reducing the signal noise by averaging a number of measures. This could be important to reduce some random errors in the solution.
- A computation hold period ΔT_{h3} for solving the next step at the integration algorithm. This period includes also the transmission time if the system equations are solved in a different CPU than the controller itself.
- A period of ramp ΔT_{ram} at the reference signal in order to smoothly change to the new computed displacement $\mathbf{d}^{n+1}$. During the three previous hold periods, the reference was maintained constant at the previous value of the computed displacement $\mathbf{d}^{n}$.

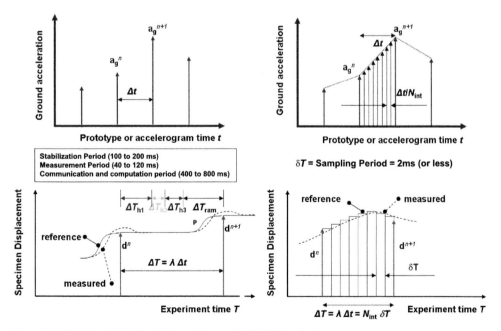

Figure 2.   The classical (left) and the continuous (right) PsD method.

The accuracy in the imposed displacement and the measured force depends, apart from the characteristics of the experimental set-up, on the selected periods for stabilising, measuring and ramp, whereas the computation period is determined by the system equations and processor characteristics. The experimental duration of one prototype time step can vary, depending on the value of the increment of displacement to apply. The asynchronous nature of the classical PsD method is a particular advantage when considering substructuring and/or distributed testing, since the delay in the computation and/or communications are already included in ΔT_{h3}. It can also be appropriate for tests that require complex transformations of coordinates at every step, as for bi-directional testing (Molina et al., 1999).

In the continuous PsD method (Figure 2 right), in contrast with the classical one, the execution of every integration step takes just one sampling period (δT) of the digital controller of the control system, e.g. 2 ms in the ELSA implementation (Magonette et al., 1998). The ramp and stabilising periods are reduced to zero duration and the measuring plus the computation periods must be feasible within those few milliseconds of experimental time. The surprising fact is that, since the hydraulic control system is unable to respond significantly at frequencies in the range of the sampling frequency of its controller, e.g. 500 Hz, the missing periods of ramp and stabilising are not needed at all.

Under these conditions, the accuracy in the imposed displacement depends basically on the testing speed, which is characterised by the time scale factor λ. In order to reduce the testing speed while the experimental time step is kept fixed to δT, every original time increment in the prototype domain is subdivided into a number internal steps N_{int} (Figure 3 right), so that, by increasing this number, the time scale factor is enlarged as: $\lambda = T/t = (N_{int}\delta T)/\Delta t$.

As shown in Figure 2 (right), at every Δt the required internal values of the input ground accelerogram are linearly interpolated from the original record values. The total number of integration steps in a continuous PsD test can be of several millions, in comparison with several thousands as typically required for a classical PsD test for the same specified earthquake. This fact implies that, on the one hand, an explicit time integration algorithm can always be used without concern about stability or integration error and, on the other hand, there is no longer need for an averaging period at the measuring of the force since any high-frequency noise at the load cells will automatically be filtered out in the solution. Such filtering effect is due to the equation response characteristics for the frequency associated to such small time increment.

Additionally, working with the continuous PsD method, it is usually possible to perform the test in a shorter experimental time, but with a better accuracy than with the classical method for the

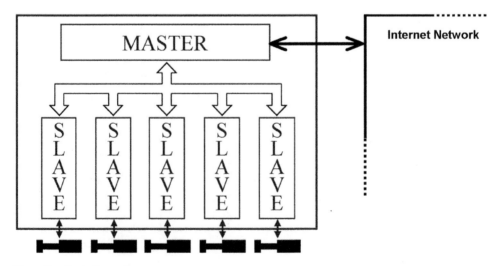

Figure 3.   Controller parts.

same experimental hardware. This is because, as a consequence of the mentioned characteristics of the continuous version, the absence of alternation between ramp and hold periods in the controller reference signal notably improves the control quality. Since all the computations are to be performed in a synchronized way within the very short sampling period δT, substructuring involving huge computations (large number of degrees of freedom or non-linearities) or distributed testing are very challenging. As a matter of fact, existing hardware is unable to cope with this kind situation so that different substructuring algorithms need to be introduced.

3   HARDWARE AND SOFTWARE AT ELSA

The implementation of the continuous PsD method has been achieved by providing substantial modifications in the hardware architecture of the testing system. The challenge is to synchronise and complete inside the control sampling time (typically 1 or 2 ms), the main tasks of a PsD cycle: measurement, motion computation and displacement control. An advanced hardware configuration has been set up to ensure a strong coupling and a very high-speed data communication between the servo-controllers and the main computer solving the equations of the motion.

In practice, the hardware consists of three main parts as described in Figure 3 (Buchet, 2006a): the master card, the slave cards (they are usually more than one) and the passive bus connection. The master card contains the kernel of the pseudo-dynamic algorithm. For this reason it is equipped with a fast processor (Pentium class) and enough memory in order to store the necessary data.

The slave card consists of a main board equipped with a dual port ram and three main components plugged: a PC104 central processing unit card, a controller signal input and output card and an analogue input/output card (see Figure 4); The ISA passive bus connects the master and slaves cards. In the current configuration it can connect up to one master card and seven slave's cards, but in principle the limitation on the number of cards that can be connected with a passive bus is 16; The assembled controller must then be put into a rack and the peripherals (USB drive, LCD screens and so on) connected to it. It is possible to reset either the master CPU or the slaves CPUs separately.

The control software (Buchet, 2006b) reflects the architecture of the hardware: there is one master program that communicates with several slaves programs. For the sake of simplicity, the following description will be made for only one slave, but it can be easily generalised to several slaves. Both the master and the slave programs originate two main processes: the background process and the foreground process. The first one is devoted to manage several services that are used during control such as the keyboard, the uploading of control parameters, the displays refresh, the hard disk management, the LAN connection, the remote services (under NT platform), the

Figure 4.   View of the controller slave board.

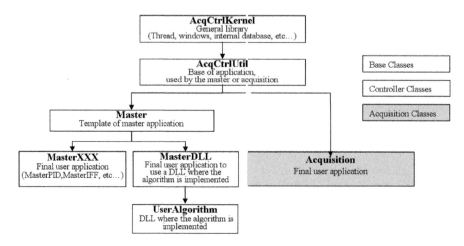

Figure 5.   Real-time application organization.

data exchange between master and slave and the data exchange between master and remote station. Since these services are not strictly necessary (the display refresh, for example, can be delayed a little bit, if it is needed), these processes have a lower priority than those in the fore-ground process.

The foreground process is the core of the control software: it performs at a fixed sample rate the data acquisition and the computation of the control variables. For this reason it must have absolute priority over the background processes because, obviously, delays cannot be accepted in the control algorithm.

The master application is a multi-thread application; the threads in background are dedicated to the user interface (graph, console), the asynchronous data exchange with the Controller and the NT station and the data acquisition on the hard disk. An interrupt generated by the specific board give the control to an interrupt routine where the pseudo-dynamic algorithm is executed and also the synchronous data exchange with the controller. The controller master support also acquisition and generator features.

The C++ software is composed of several modules following the hierarchy of Figure 5.

The UserAlgorithm DLL gives the possibility to the final users to write their own algorithm without recompile and touch the application Master.exe. The DLL can be used by Matlab to test the algorithm in mode offline and by the application Master.exe in mode online (real-time). The openness of the system allows using it for any multi-actuator configurations such as PsD testing, cyclic & fatigue testing or structural control implementation.

4   CONTINUOUS TESTING WITH SUBSTRUCTURING: MONOLITHIC AND DISTRIBUTED APPROACHES

Monolithic substructure technique means that the process running at the level of the master (and implementing the masterDLL algorithm) handles all the DoFs of the structure, analytical and experimental. Since it is not desirable to work with heavy complex algorithms at the level of masterDLL (all the operations need to be performed as fast as possible in order to let the back-ground processes a chance to perform their job), this approach is reserved to the case of simple structures involving an elastic analytical part that can be represented by a small size mass, damping and stiffness matrices.

It is thus well adapted for subcomponent testing and in particular the one dealing with isolators. Using the *same* testing device (usually a simple 1 DoF system), it is possible to characterize the

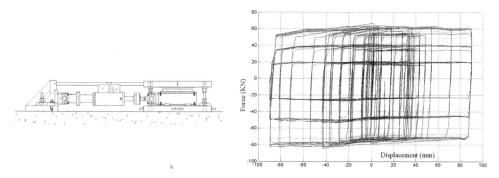

Figure 6.   UHYDE testing setup (left) and characterization of the friction device (right, from Bossi 2003).

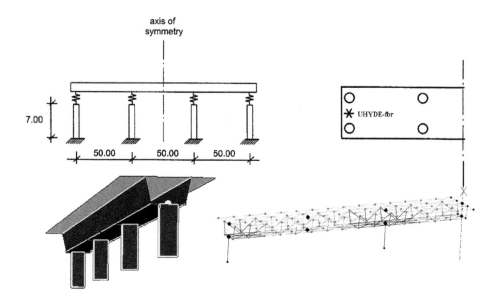

Figure 7.   The bridge configuration.

isolation system by imposing to it a predefined loading history *and* testing it in realistic seismic conditions combining it with a numerical substructure.

This situation is illustrated considering the setup of the left part of Figure 6. A special dissipation device (UHYDE-fbr) is put between two fixed steel plates. The lower part of the device is free to slide on a Teflon plate whereas the upper part is rigidly connected to the upper steel plate. Four actuators put at the top of the upper plate can impose a constant pressure to the membrane.

The results of the characterization of the devices are shown in the right part of Figure 6. A cyclic loading has been imposed with an internal pressure of 2 (blue), 4 (red), 6 (green) and 7 (black) bars. This almost elastoplastic behaviour is not modified by the velocity of the loading.

Now this device is used to isolate a structure, in this case the bridge of Figure 7. The structure (and the use of the isolators) being symmetrical, only halve of the structure has been considered. Anticipating the use of isolators working mainly in shear at the top of the piers, elastomeric supports have been considered in order to support the weight of the deck in static condition and to re-align the deck after a seismic event. The structure (and the elastomeric support) has been modelled using finite elements (beams and plates), further statically condensed on 20 carefully selected DoFs. Thus the analytical substructure is characterized by its stiffness **K**, damping **C** and mass **M** matrices.

Only the configuration A, where the UHYDE-fbr has been mounted at the top of the external piers is considered here.

Assuming that r_{iso} is the value of the actuator force measured by the load cell at the level of the slave controller, the central difference algorithm, implemented within the MasterDLL application and running in foreground in the master, just has to assemble the elastically modelled response of the bridge with the force vector constructed from r_{iso}.

A typical seismic response in displacement is given in Figure 8 left, with 10 cm of amplitude (30% of the seismic input, 6 bars pressure on the membrane). Several curves have been plotted, each of them obtained with a different value of λ, the time scale factor (blue: $\lambda = 5$, red: $\lambda = 3$, green: $\lambda = 2$ and black: $\lambda = 1$). It is thus possible to perform real time subcomponent tests (case with $\lambda = 1$). However the displacement error increases almost quadratically when raising the speed of execution of the experiment, as illustrated in Figure 8 right where the plot of the logarithm of a measure of this error with respect to the logarithm of λ exhibits a slope close to 2.

The monolithic approach is however restricted to the case of simple elastic analytical structure and the extension of the substructuring techniques to the general case introduces some challenging difficulties, namely:

- If the analytical part of the structure is complex, even an elastic computation may become unfeasible within a control period of the experimental process. Two processes are thus likely to be used, the experimental one on the master controller, and the analytical one, based on finite element modelling and running on a remote workstation.
- It is possible to consider the experimental part as a special element of the model. In this approach, relying on the classical PsD approach, the trajectory of the experimental DoFs is assumed and no longer related to time integration. In absence of robust alternative, this approach has been considered for the large bridges tests performed in 2001 to investigate vulnerability issues (Pinto et al., 2004).
- In order to preserve the essence of the continuous PsD approach (time integration involving all the force measurements), the time integration for the experimental and the modelled parts of the structure have to be performed with different time steps.
- The experimental process works usually with the explicit central difference scheme, thus using small time steps. Again, when the analytical structure is complex, it is not evident that using the same explicit scheme would allow to obtain stable results. Thus different time integration schemes are likely to be used by the two processes.
- The communication between the two processes could not be a priori staggered. Since the experimental process is synchronized, it is not desirable to stop it systematically in order to wait for information coming from the analytical process. What is desirable is that the experimental process could read (and send) information from (and to) the analytical process at some prescribed instant and then proceed without stopping.

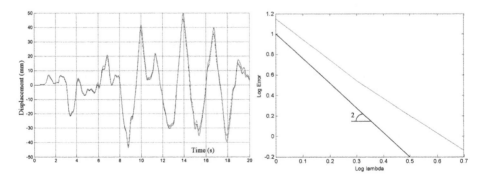

Figure 8.   Seismic response of the isolator (left) and error convergence with λ (right).

These difficulties have been overcome recently relying on the algorithm presented by Gravouil & Combescure, 2001 and transforming this essentially staggered asynchronous procedure in an inter-field parallel procedure, suitable to work with synchronous processes (Pegon & Magonette, 2002), and non-linear modelling (Pegon & Magonette, 2005).

In this method, the structure under consideration is split into two subdomains A and B. The continuity of the velocity is assumed between the two domains and ensured using a Lagrange multiplier technique. Equation 1 thus becomes

$$\mathbf{M}_A\mathbf{a}_A + \mathbf{C}_A\mathbf{v}_A + \mathbf{r}_A(\mathbf{d}_A) = \mathbf{f}_A + {}^T\mathbf{L}_A\Lambda \; \mathbf{M}_B\mathbf{a}_B + \mathbf{r}_B(\mathbf{d}_B)$$

$$= \mathbf{f}_B + {}^T\mathbf{L}_B\Lambda \; \mathbf{L}_A\mathbf{v}_A + \mathbf{L}_B\mathbf{v}_B = 0 \qquad (2)$$

where $\mathbf{L}_A$ and $\mathbf{L}_B$ are connectivity matrices expressing a linear relationships between the connected boundaries of subdomains A and B, and Λ the vector of Lagrange multipliers. Note that this formulation put into duality the connected cinematic quantities (velocities) and the resulting reaction forces (Lagrange multipliers) modifying the equilibrium on each subdomain.

Typically, in the context of the substructuring technique, domain A would be the analytical part of the structure, integrated in time using the trapezoidal rule, whereas domain B would be the experimental part, integrated using the central difference scheme. We can thus assume that A is associated with the coarse time scale Δt whereas B uses the fine time step $\Delta t/N_{\text{int}}$. Note that a damping matrix $\mathbf{C}_A$ has been introduced for the analytical part. Since domain A corresponds to an analytical structure, it is important to introduce a viscous equivalent damping to represent micro-hysteresis at low amplitude.

The algorithm of Gravouil & Combescure, 2001 is staggered: the domain A sends a velocity information ($\mathbf{v}^{n+1}$) related to t^{n+1}, allowing domain B to perform its substeps from t^n to t^{n+1} and sending back to A a force information (Λ^{n+1}) allowing A to proceed. Clearly B has thus to pause, waiting for A (top of Figure 9). In order to give the experimental process a chance to proceed without pausing, the velocity information has to be known in advance. A modification of the scheme, using two interlaced time integration schemes with a double time step ($2\Delta t$) has been introduced (Pegon & Magonette, 2002) as illustrated in the bottom of Figure 9. In this modification

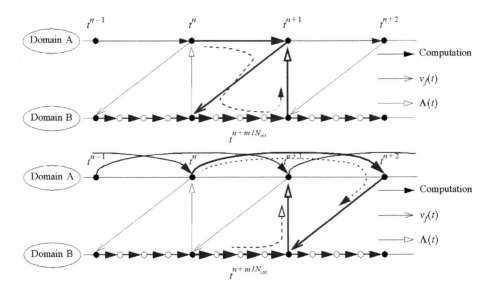

Figure 9.　Staggered (top) and inter-field (bottom) versions of the coupling procedure.

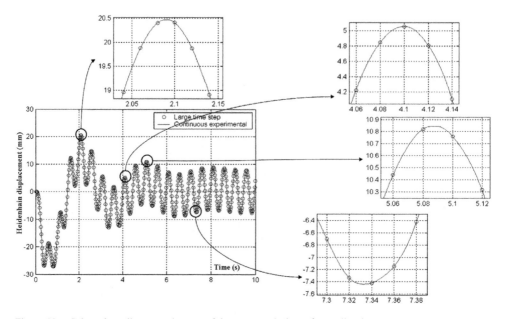

Figure 10.   Selected small steps trajectory of the actuator during a force vibration test.

the following important feature of the original scheme is maintained: the overall scheme is stable as soon as the central difference scheme running for the laboratory part is stable.

Using the UHYDE setup of Figure 6, extensive laboratory tests of the procedure has been performed. An example of the results obtained with $\lambda = 10$ and $N_{int} = 100$, in a forced vibration case, is shown on Figure 10 in order to underline the smoothness of the laboratory trajectories.

5   CONCLUDING REMARKS

Some of the main capabilities and achievements in structural testing at the ELSA laboratory have been summarised in this chapter. As a complement to other laboratories in Europe based on shaking-tables facilities, ELSA has specialised itself in tests on large-size models and with sophisticated computer-controlled load-application conditions. Internationally recognized pioneering steps have been achieved for the development of the PsD testing method and its full-scale implementation. The ELSA contribution includes also some cases of real dynamic tests of active and semi-active control systems as well as vibration monitoring.

Robust implementation of substructuring has always been an ELSA priority, leading to pioneering tests on bridges. The current development effort is put on using sophisticated domain decomposition techniques able to preserve the smooth character of the continuous PsD testing, and, at the same time, able to adapt with foreseen hardware developments (actuators, control, CPU, network), to tend versus reliable and high quality real time testing.

REFERENCES

Bossi, A. 2003. Isolamento sismico: prove di dispositivi dissipativi con il metodo pseudo-dinamico continuo con sottostrutturazione, *Tesi di Laurea*, Universita degli studi di Perugia, Italy.
Buchet, Ph. 2006a. Controller Users Manual. *EUR-Report no. 22329 EN*, Ispra, Italy.
Buchet, Ph. 2006b. Acquisition & Control Software Architecture. *EUR-Report no. 22128 EN*, Ispra, Italy.

Combescure, D. & Pegon, P. 1997. α-Operator Splitting time integration technique for pseudodynamic testing. Error propagation analysis. *Soil Dynamics and Earthquake Engineering* 16: 427–443.

Casciati, F., Magonette, G., Marazzi, F. 2006. *Semiactive Devices and Applications in Vibration Mitigation*. Chichester: John Wiley & Sons Ltd.

De Luca, A., Mele, E., Molina, F.J., Verzeletti, G., Pinto, A.V. 2001. Base isolation for retrofitting historic buildings: evaluation of seismic performance through experimental investigation. *Earthquake Engineering & Structural Dynamics* 30: 1125–1145.

Donea, J., Magonette, G., Negro, P., Pegon, P., Pinto, A., Verzeletti, G. 1996. Pseudodynamic capabilities of the ELSA laboratory for earthquake testing of large structures. *Earthquake Spectra* 12: 163–180.

Gravouil, A. & Combescure, A. 2001. Multi-time-step explicit-implicit method for non-linear structural dynamics. *Int. J. Numer. Meth. Engng.* 50: 199–225.

Magonette, G., Marazzi, F. 2001. Active Control of Civil Structures: Theoretical and Experimental Study. *IMAC-XIX: Conference on Structural Dynamics*, February 5–8, Hyatt Orlando, Kissimmee, Florida, Society for Experimental Mechanics, Inc., ISBN: 0-912053-72-0: 524–529.

Magonette, G., Pegon, P., Molina, F.J., Buchet, Ph. 1998. Development of fast continuous substructuring tests. *Proceedings of the 2nd World Conference on Structural Control*, Session TD01d, p. 69.

Molina, F.J., Verzeletti, G., Magonette, G., Buchet, Ph., Géradin, M. 1999. Bi-directional pseudodynamic test of a full-size three-storey building, *Earthquake Engineering & Structural Dynamics* 28: 1541–1566.

Molina, F.J., Magonette, G., Pegon, P. 2002a. Assessment of systematic experimental errors in pseudodynamic tests. *Proc. of 12th European Conference on Earthquake Engineering*. Paper 525. Oxford: Elsevier Science Ltd.

Molina, F.J., Verzeletti, G., Magonette, G., Buchet, Ph., Renda, V., Geradin, M., Parducci, A, Mezzi, M, Pacchiarotti, A, Federici, L., Mascelloni, S. 2002b. Pseudodynamic tests on rubber base isolators with numerical substructuring of the superstructure and strain-rate effect compensation. *Earthquake Engineering & Structural Dynamics* 31: 1563–1582.

Molina, F.J., Sorace, S., Terenzi, G., Magonette, G., Viaccoz, B. 2004. Seismic tests on reinforced concrete and steel frames retrofitted with dissipative braces. *Earthquake Engineering & Structural Dynamics* 33: 1373–1394.

Molina, F.J. & Géradin, M., 2007, Earthquake Engineering experimental Research at JRC-ELSA. NATO workshop. Extreme Man-Made and Natural Hazards in Dynamics of Structures (NATO Security through Science Series C: Environmental Security), Ibrahimbegovic, A. & Kozar, I. (Editors), Springer, 311–351.

Pegon, P. & Magonette, G. 2002. Continuous PsD testing with non-linear substructuring: presentation of a stable parallel inter-field procedure. *JRC-Special publication no. SPI.02.167*, Ispra, Italy.

Pegon, P. & Magonette, G. 2005. Continuous PsD testing with non-linear substructuring: using the Operator Splitting technique to avoid iterative procedures. *JRC-Special publication no. SPI.05.30*, Ispra, Italy.

Pinto, A.V., Pegon, P., Magonette, G., Tsionis, G. 2004. Pseudo-dynamic testing of bridges using non-linear substructuring. *Earthquake Engineering & Structural Dynamics* 33: 1125–1146.

Pinto, A.V., Pegon, P., Taucer, F. 2006. Shaking table facilities and testing for advancement of earthquake engineering: international cooperation, experiences, values, chances. *First European Conference on Earthquake Engineering and Seismology* Paper 1538, Geneva, Switzerland.

Shing, P.B., Mahin, S.A. 1987. Cumulative experimental errors in pseudo-dynamic tests. *Earthquake Engineering & Structural Dynamics* 15: 409–424.

Takanashi, K., Nakashima, M. 1986. A State of the Art: Japanese Activities on on-Line Computer Test Control Method. *Report of the Institute of Industrial Science* 32, 3, University of Tokyo.

CHAPTER 9

Hybrid testing facilities in Korea

Chul-Young Kim
Dept. of Civil & Environmental Engineering, Myongji University, South Korea

Young-Suk Park
Dept. of Civil & Environmental Engineering, Myongji University, South Korea

Jae-Kwan Kim
School of Civil, Urban & Geosystem Engineering, Seoul National University, South Korea

ABSTRACT: In 2004, The Korea Ministry of Construction and Transportation (MOCT) launched the Korea Construction Engineering Development Collaboratory Program (KOCED Program) to establish a comprehensive base for construction-related testing, research and education. With the ultimate goal of strengthening Korea's international competitiveness in construction technologies, the KOCED Program aims to promote research and development and to set up a nationwide education program to produce highly qualified researchers and practitioners in the various fields of construction engineering. During the next decade 12 large scale experimental facilities, two of which are hybrid structural testing facilities, will be built and operated at the major regional universities. These facilities are going to be linked with the users along with a digital data repository and supercomputers using a grid architecture high performance information network. This paper outlines the KOCED Collaboratory Program and reports the current progress and future plans. In addition, among the 6 KOCED testing facilities, a real-time hybrid structural testing facility (HySTeC) which are currently under construction at Myongji University, is introduced briefly.

1 KOCED AND TESTING FACILITIES

1.1 *KOCED program*

The Korean government has invested a large amount of budget in the construction of social infra structure. New highways, airports, high speed railroads and long span bridges have recently been built. Modern high rise buildings and numerous apartment complexes were also constructed. But the massive construction work failed to be linked to the progress being made in the design and construction technologies. One of the causes is the lack of large scale testing facilities, especially in the structural engineering field. The advancement and development of state-of-the-arts technologies and original design methods requires good experimental facilities. The research community felt very strongly that large scale experimental research facilities should be built with urgency in order to be competitive globally, and to enhance the design and construction quality domestically. The civil engineering profession in Korea was in need of renovation and overhaul.

In order to achieve this objective efficiently and economically, a new ambitious project, named as the Korea Construction Engineering Development Collaboratory Program (KOCED Program) was launched by the Korea Ministry of Construction and Transportation. Its objective is to build 12 large scale testing facilities at the major universities evenly distributed around the country and interconnect them using a high performance information network. It will become a collaboratory, operating on a shared-use basis. A digital data repository and a high performance computing facility will be integrated into this grid system. It was inspired by the NEES Program of the United States. While NEES is concerned with earthquake engineering only, KOCED Program encompasses the entire civil engineering field.

In 2004 this program was officially started. As the first step, the Korea Ministry of Construction and Transportation (MOCT) created the KOCED Program Management Center (KOCED PMC) for the purpose of managing the whole program. The mission of the KOCED PMC is to develop a consortium that will operate the collaboratory, to develop and implement grid system that will inter-link all the facilities, and to manage the construction process of the testing facilities. Because Korea was starting from scratch, the KOCED program proceeded in a top-down format. The plan was written by a group of dedicated investigators in 2003 with the following objectives in mind:

- At least one experimental facility for each discipline of construction engineering is going to be built.
- All the facilities should be brand new and up-to-date. Their capacity and performance should meet international standards.
- The whole collaboratory should be shared by the entire construction engineering community of Korea.
- The facilities should be utilized for both research and educational purposes.
- The locations of the facilities should be evenly distributed around the country in order to contribute to the balanced development of the country.
- The Program should accommodate and promote synergy with leading edge technologies in order to renovate the outdated construction engineering standards.

It became evident that the above objectives could be satisfied only if the experimental facilities are tied together using a high performance information network and the collaboratory is operated by a consortium as demonstrated by the NEES Program. Therefore the NEES model was adopted for the KOCED Program. This concept is described in Figure 1. In the KOCED Collaboratory, however, the experimental facilities are not limited to earthquake engineering applications, but extend over the whole civil and construction engineering applications.

The original time table of this project is shown in Figure 2. According to this plan, 12 experimental facilities will be built, the grid system called "KOCED grid" will be constructed, and the consortium will be developed for 6 years from 2004 through 2009. The budget for the construction of 12 facilities and development of the KOCED grid and KOCED Consortium is expected to exceed

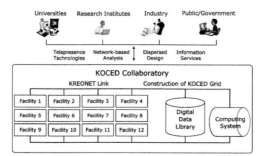

Figure 1.   Concept of KOCED collaboratory.

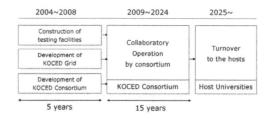

Figure 2.   Strategic plan of the KOCED program.

100 million USD. During the 15 year period starting from 2009, the KOCED Collaboratory will be operated on a shared-use basis and managed by the KOCED Consortium. In the year 2025, all the facilities will be turned over to the hosting universities.

1.2   KOCED testing facilities

The 12 experimental facilities and their capacity and performance, as they were planned initially, are briefly listed in Table 1.

Table 1.   Dimensions and capacities of 12 testing facilities.

Stage	Facility	Dimension and capacity
1	Real time hybrid structural testing facility	Dynamic Actuators: 250 kN, 1000 kN, 2000 kN Reaction Walls: 21 m × 12 m × 3 m, 12 m × 12 m × 3 m
	Geo-centrifuge	Geotechnical Centrifuge: 5 m (Radius), 240 g · ton 2D Shaking Table, In-flight Robot
	Multi-platform seismic simulation facility	One Fixed Shaking Table (2 DOF): Size: 5 m × 5 m; Two Movable Shaking Table (2 DOF): Size: 3 m × 3 m
	Advanced construction materials testing facility	5 MN Concrete Compression Testing machine 600 kN UTM (Steel), 100 kN UTM (Composite Materials)
	Large boundary layer wind tunnel	Two Story Wind tunnel: 10 m × 2 m × 10 m, 0.5–10 m/sec
	Ocean environment simulation wave tanks	3D Irregular Type Wave Generator: 50 m × 50 m × 1.5 m 2D Irregular Type Wave Generator: 90 m × 2 m × 1.5 m
2	Extreme load testing facility	Impact & Collision Test Equipment Explosion Chamber
	Test facility for long-term environmental effects	Durability Test Equipment: EPMA, SEM, XRF, XRD Creep, Shrinkage Test Equipment
	Multi-purpose field test facility	Static Actuators: 300 ton, Shaker: 0.2 Hz~15.0 Hz Soil Box: 10 m × 10 m × 3 m
	Mobile geotechnical laboratory	Mobile Test Equipment for Site Investigation High Speed Wireless Network
	Reconfigurable large-scale structural testing facility	Dynamic Actuators: 500 kN, 1000 kN Fixed Reaction Walls: 16 m × 5 m × 12.25 m
	Large-scale hydro-model test basin	Open Channel: 50 m × 1.2 m × 2 m Circulation Channel: 18 m × 12 m

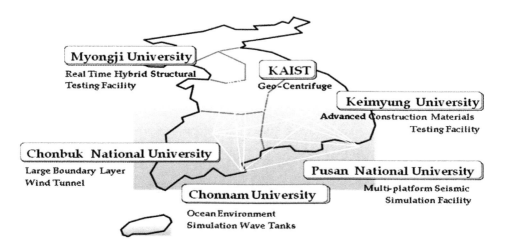

Figure 3.   Hosts of the first 6 testing facilities.

In 2004 the KOCED PMC completed the basic design for the 12 facilities. Hosting universities for the first 6 facilities were determined through open competition in December 2004 as shown in Figure 3. A research center was established at each hosting universities to manage its testing facility. Detail designs of these facilities have been completed followed by the actual design starting from August 2006. All the six centers have held the ground breaking ceremony and they are under construction. It is expected that the construction will be finished by mid 2008 and after the test operation for a couple of months they will be opened to public service from the beginning of 2009.

2   HYBRID STRUCTURAL TESTING CENTER (HySTeC)

2.1   *Introduction*

Even though there have been many conventional structural testing facilities in Korea and some quasi-static and pseudo-dynamic tests have been carried out, HySTeC of Myongji University will become the first one which can perform real-time hybrid structural tests. Like the other KOCED testing facilities, HySTeC is currently under construction and is expected to be completed by mid 2008 and to be open to public service since the beginning of 2009. Total of 11,000,000 USD (8 million from MOCT and 3 million from Myongji University) will be spent mostly for construction of the building and purchase of testing equipments.

It has 38.4 m × 12.8 m main reaction floor with 10 m × 9 m outdoor extended strong floor so that it can accommodate several tests simultaneously or very long test specimen up to 50 m. The reaction

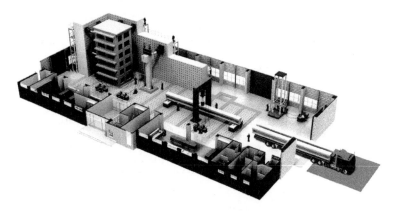

Figure 4.   Rendering of HySTeC.

Figure 5.   Various actuators.

wall is L-shaped with short side of 8 m long and long side of 24 m long and is staggered with height variation of 12 m, 9 m and 6 m as shown in Figure 1. Two overhead cranes of 300 kN capacity will carry even the large and heavy specimen with ease.

2.2 *Testing equipments and software*

One of the important H/W requirements for fast hybrid testing would be the robust hydraulic capacity in order to provide sufficient hydraulic flow compatible with strong earthquake motion. An array of four 180 gpm hydraulic pumps, total of 720 gpm, supplies hydraulic flow enough to run single hybrid or dynamic test which requires high speed actuator movement or several static and fatigue tests simultaneously. Accumulators are also under consideration for future impulsive tests which requires very high flow rate in short period (Elnashai et al., 2004; MTS 2005; Stojadinovic et al., 2004).

Even though the real-time hybrid testing is the most important function that should be implemented in this facility, it is expected that a large portion of the testing needs would be conventional static, fatigue or dynamic tests requiring various actuating load capacities, speeds and strokes. To accomplish these diverse testing needs, a series of actuators ranging from 250 kN to 5,000 kN are equipped for the first stage. It is worth mentioned that 250 kN, 1,000 kN and 2,000 kN actuators adopt dual servo valves, small capacity for static tests which requires very small hydraulic flow control and large capacity for dynamic tests which requires high hydraulic pressure. And also, every actuator has a long stroke enough to make ductile specimens to reach failure status as well as a high speed to simulate real-time earthquake motion. Specifications of the actuators are shown in Table 2.

In addition to these actuators, 5 MN static UTM mainly for concrete material test and 5 MN dynamic UTM for both static and fatigue tests of structural members as well as material specimens.

More than 450 channels of static signals can be measured and stored by high-speed static data loggers with 1000 ch/sec scanning rate and a dynamic DAQ system is to acquire 128 channels of dynamic signals with up to 100 kHz sampling rate per each channel. This DAQ system also has the shared memory interface to communicate with other hybrid testing equipments for fast data transfer.

Two 4-channel FlexTest GT controllers are capable of controlling up to 4 channels of actuators or up to 4 independent test stations depending on laboratory testing needs. In addition to conventional structural control capabilities, one of the controllers is configured with additional hardware such as the shared memory interface to allow it to be used for a real-time hybrid simulation. They also include pseudo-dynamic testing, quasi-static step-by-step testing and seismic waveform generation software and multi-purpose testing software as well as simulation interface to Berkeley's OpenSee's and OpenFresco hybrid simulation software (Mazzoni et al., 2006). In addition to these two 4-channel controllers, one 2-channel FlexTest GT controller and one single channel FlexTest SE controller are equipped for general purpose tests that should be carried out simultaneously.

Mathwork's Simulink tools are integrated into controller platform to achieve real-time hybrid control capabilities (Mosqueda, 2004). The control architecture consists of several primary components:

1. Digital servo controller which performs all aspects of test set-up, hydraulic flow management, servo control and data acquisition;
2. A host PC for running MATLAB®, Simulink®, Real-Time Workshop, x-PC Target and a C compiler to create your real-time applications;
3. A Target PC for running x-PC Target real-time kernel for interfacing the digital servo controller;
4. Shared memory functionality between the Target PC and the controller;
5. Software tools for managing the shared memory locations for feedback and command signals in real time between the x-PC Target computer and the controller and for interfacing the dedicated real time DSP managing test interaction between the real test article and the model.

A schematic diagram of the above concept is shown in Figure 6.

Table 2.   Specifications of actuators.

Load capacity [kN]	Number	Servo valve [gpm]	Velocity [mm/sec]	Stroke [mm]	Remark
250	2 ea.	15/180	896	750	Dual
1,000	2 ea.	15/250	323	750	Dual
2,000	2 ea.	15/250	161	1000	Dual
5,000	2 ea.	180	–	1000	Single

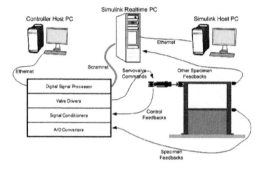

Figure 6.   Conceptual drawing of hybrid control flow.

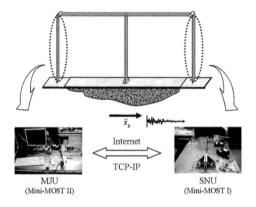

Figure 7.   Distributed mini hybrid test.

2.3   *Research activities*

During the construction of the center, a study on the hybrid simulation technique has started with a mini model repeating the MiniMOST system of NEES (Kwon et al., 2005; Yang et al., 2004). Original MiniMOST system that is built to verify the feasibility of hybrid simulation algorithm uses NEESPop server for communication, MatLAB for numerical simulation and a linear stepping motor for actuation so that it has a very slow control speed. Modification and improvement have been carried out on this original system by integrating the direct communication with both the controller and the DAQ. And also modified OpenFresco and OpenSees software were migrated into this system and the distributed hybrid test through the internet has been successfully carried out between Myongji University(MJU) and Seoul National University(SNU). Figures 4 and 5 show the test scheme and a typical result where numerical and hybrid simulation results at both sites are almost identical.

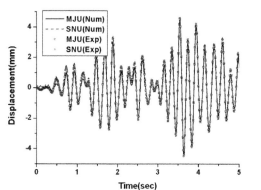

Figure 8.   Results of distributed hybrid test.

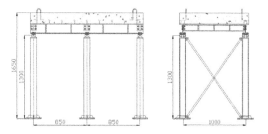

Figure 9.   A model for benchmarking shaking table test.

Further research is going on to build another small-scale system with small hydraulic actuators and a FlexTest GT controller that will be used for the actual large-scale test. Local hybrid simulations with substructures as well as a benchmarking shaking table test of a 2-bay 1-story full model in Figure 9 will be performed. From this system and result, it is expected that previous hybrid simulation algorithms can be compared and verified.

3   CONCLUDING REMARKS

The main objective of the KOCED program is to lift the technology standards of the civil and construction engineering fields to a higher level. The outcome of this program will be welcoming, efficient and safe infra-systems. The benefactors will be the public. Moreover, the benefits should not be limited to within the borders of Korea, but should be shared by the world community. KOCED community including the two hybrid testing facilities will be willing to share facilities, information and technology with other researchers in the world to enhance overall construction technology as well as the hybrid simulation technique itself.

REFERENCES

Elnashai, A., Spencer, B., Kuchma, D., Ghaboussi, J., Hashash, Y. & Gan, Q. 2004. Multi-axial Full-scale Sub-structured Testing and Simulation (MUST-SIM) facility at the University of Illinois at Urbana-Champaign. *13th World Conference on Earthquake Engineering*. 1–6 August 2004, Vancouver, B.C., Canada.
Kwon, O., Nakata, N. Elnashai, A. & Spencer, B. 2005. technical note: A Framework for Multi-site Distributed Simulation and Application to Complex Structural Systems. *Journal of Earthquake Engineering* 9(5): 741–753.

Mazzoni, S., McKenna F., Scott, M.H. & Fences, G.L. 2006. *OpenSees Command Language Manual*. PEERC. University of California, Berkeley.

Mosqueda, G. 2004. *User's Manual-Hybrid Controller*. Department of Civil and Environmental Engineering, University of California, Berkeley.

MTS, 2005. *Civil, Structural and Architectural Engineering Testing Capabilities*. MTS Systems Corporation.

Stojadinovic, B., Moehle, J.P., Mahin, S.A., Mosalam, K. & Canny. J.F. 2004. Nees Equipment Site at the University of California, Berkeley. *13th World Conference on Earthquake Engineering*. 1–6 August 2004, Vancouver, B.C., Canada.

Yang, G., Spencer, B. & Myers, M. 2004. *Instrument Wiring and Setting for Mini-MOST Experiment*. UIUC MUST-SIM Facility. Urbana, IL.

CHAPTER 10

NEES@Lehigh: Real-time hybrid pseudodynamic testing of large-scale structures

O. Mercan & J.M. Ricles
Lehigh University, Bethlehem, PA, USA

ABSTRACT: The real-time hybrid pseudodynamic (PSD) test method, by combining computer simulation and physical testing, offers a viable means of investigating the behavior of rate-dependent structures under dynamic loading. The implementation and use of the real-time hybrid PSD test method at the Lehigh NEES equipment site is presented. The servo-hydraulic system components, integrated control hardware and software, and algorithms for real-time testing at the Lehigh NEES equipment site are described. Several requirements have to be met to perform a successful real-time test. The stability boundary associated with restoring force delays in analytical and experimental substructures in a real-time test setup and a velocity feed forward compensation method to reduce actuator delay in the experimental substructure are discussed. Results from MDOF real-time hybrid PSD tests are presented and evaluated by means of tracking indicators. The application of the real-time hybrid PSD test method in determining the requirements for passive dampers to enable a structure to meet seismic performance-based design objectives is presented.

1   INTRODUCTION

The PSD test method is an experimental technique for investigating the dynamic response of structures. During a PSD test, an integration algorithm is used to directly solve the equations of motion to obtain command displacements. These command displacements are then imposed on a test structure by hydraulic actuators. The displacements as well as the restoring forces developed in the deformed test structure are measured and fed back to the integration algorithm, and are subsequently used in the generation of the command displacements for the next time step. The hybrid PSD test method (Dermitzakis & Mahin, 1985) is a practical extension of the PSD test method, where the structure is separated into two parts. The component of the structure for which a reliable analytical model does not exist is tested physically in the laboratory and called the experimental substructure. The remaining part of the structure is modeled in a computer, and is called the analytical substructure. During a hybrid PSD test, the command displacements generated by the integration algorithm are simultaneously imposed to the experimental and analytical substructures, and the corresponding restoring forces from both are combined and fed back to the integration algorithm.

When the experimental substructure has load-rate dependent properties, it becomes necessary to perform the test at fast rates (ideally in real time) to accurately capture the structural response. Real-time testing requires accurate control of servo-hydraulic actuators with synchronized communication between an inner loop (consisting of the combined servo-hydraulic system and test structure) and an outer loop (consisting of the combined integration, actuator delay and kinematic compensation algorithms, and analytical substructure for a hybrid test). Errors in the restoring force from the experimental or analytical substructures, especially a time delay, have detrimental effects and must be minimized to maintain the accuracy and stability of the test. This chapter contains information related to the implementation of the real-time hybrid PSD test method at the Lehigh NEES equipment site, and discusses important aspects for achieving a successful real-time test.

2   REAL TIME HYBRID PSD TESTING

2.1   *NEES RTMD facility*

The Real-time Multi-directional (RTMD) Earthquake Simulation Facility was been established in 2004 at the Lehigh University ATLSS Engineering Research Center, and is an equipment site within the Network for Earthquake Engineering Simulation (NEES). The ATLSS Laboratory features a strong floor that measures 31.1 m by 15.2 m in plan, and a multi-directional reaction wall that measures up to 15.2 m in height (see Figure 1). Anchor points are spaced on a 1.5 m grid along the floor and walls. Each anchor point can resist 1.33 MN tension force and 2.22 MN shear force. Additional steel framing is used in combination with the strong floor and reaction wall to create a wide variety of test configurations.

To create the RTMD facility, several pieces of equipment have been installed in the ATLSS Laboratory. This equipment includes five dynamic, double rodded hydraulic actuators with a +/−500 mm stroke. Two of these actuators have a 2300 kN maximum load capacity, with the remaining three having 1700 kN maximum load capacity. Each of the actuators is ported for three 1500 liters/min servo-valves, enabling them to achieve a maximum nominal velocity of 840 mm/sec (2300 kN actuators) and 1140 mm/sec (1700 kN actuators). The existing hydraulic power supply system at ATLSS consisted of five 2250 liter/min pumps. A hydraulic oil reserve and two banks of accumulators were added to enable strong ground motion effects to be sustained for up to 30 seconds. The accumulators supply a total accumulated oil volume of 3030 liters.

Figure 1.   RTMD facility multidirectional reaction wall and strong floor.

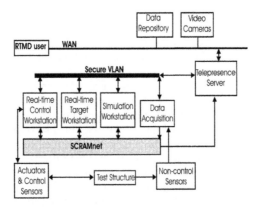

Figure 2.   Lehigh RTMD integrated control system architecture.

The servo-hydraulic integrated control system architecture for real-time testing at the RTMD facility is shown in Figure 2. An 8-channel digital controller with a 1024 Hz clock speed (identified as the Real-Time Control Workstation in Figure 2), controls the motion of the actuators through a closed servo-control loop. The Real-Time Control Workstation is integrated with the Simulation Workstation, Real-Time Target Workstation, Data Acquisition Mainframe, and the Real-Time Telepresence Workstation using SCRAMNet. SCRAMNet is a fiber optic communication device that enables shared memory and time synchronization to the Target Workstation. The Target Workstation communicates with the Control Workstation and Data Acquisition Mainframe using SCRAMNet, thereby providing a single synchronization for real-time testing. The Data Acquisition Mainframe controls a high speed 256-channel system capable of acquiring data at 1024 Hz per channel. The integrated control system architecture permits complex testing algorithms, servo-hydraulic control laws, actuator delay compensation algorithms, multi-directional kinematic compensation algorithms, and analytical substructures to be developed on the Simulation Workstation and downloaded onto the Target Workstation using SIMULINK (2007) and Mathworks xPC Target software.

The integrated control system promotes teleobservation through a Telepresence Server which links networked digital video and synchronized SCRAMNet data. The digital video is acquired from pan-tilt-zoom web cameras and fixed position cameras that are controlled through a user interface on the Telepresence Server. Live video feeds are shared with remote users through a software system designed to enable remote viewing with robotic control of video via a web browser. Experimental data and digital video is acquired and synchronously archived from SCRAMNet. Remote users can view live and archived data.

Prior to conducting a test, the task of passing metadata to and from each workstation is performed using a secure VLAN connection.

2.2 *Real-time hybrid PSD testing algorithm*

The algorithm used at the Lehigh RTMD facility for real-time hybrid PSD testing is based on the Hilber α-method (Hilber et al., 1977). The hybrid testing implementation of the α-method implicit integration algorithm is shown in Figure 3. The equations of motion and integration scheme used in the method are:

$$\mathbf{Ma}_{i+1} + (1+\alpha)\,\mathbf{Cv}_{i+1} - \alpha\mathbf{Cv}_i + (1+\alpha)\,\mathbf{r}_{i+1} - \alpha\mathbf{r}_i = (1+\alpha)\,\mathbf{P}_{i+1} - \alpha\mathbf{P}_i \qquad (1)$$

$$\mathbf{d}_{i+1} = \mathbf{d}_i + \Delta t\mathbf{v}_i + (\Delta t)^2\,[(0.5 - \beta)\mathbf{a}_i + \beta\mathbf{a}_{i+1}] \qquad (2)$$

$$\mathbf{v}_{i+1} = \mathbf{v}_i + \Delta t[(1-\gamma)\mathbf{a}_i + \gamma\mathbf{a}_{i+1}] \qquad (3)$$

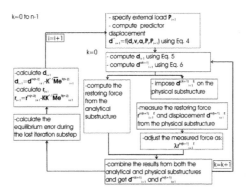

Figure 3. Modified real-time hybrid testing algorithm.

In Equation 1 $\mathbf{M}$, $\mathbf{C}$, $\mathbf{r}$, and $\mathbf{P}$ are equal to the mass matrix, viscous damping matrix, total restoring force vector, and load vector, respectively, and i and $i + 1$ are associated with the time step. The Newmark direct integration equations are used in Equation 2 to relate the displacements $\mathbf{d}_{i+1}$ at time step $i + 1$ to the displacements $\mathbf{d}_i$, velocity $\mathbf{v}_i$, and accelerations $\mathbf{a}_i$ at time step i, along with the accelerations $\mathbf{a}_{i+1}$ at time step $i + 1$, and in Equation 3 the velocity $\mathbf{v}_{i+1}$ at time step $i + 1$ to the velocity $\mathbf{v}_i$ and accelerations $\mathbf{a}_i$ at time step i, along with the acceleration $\mathbf{a}_{i+1}$ at time step $i + 1$. In the above equations Δt is the time step size and α, β and γ are integration constants. To attain unconditional stability and a favorable energy-dissipation property, Hilber et al. (1977) recommended that $\beta = (1 - \alpha)^2/4$, $\gamma = 1/2 - \alpha$, and $-1/3 \leq \alpha \leq 0$. If $\mathbf{v}_{i+1}$ from Equation 3 is substituted into the equilibrium equations, the accelerations $\mathbf{a}_{i+1}$ at the end of current time step are obtained, and the displacements $\mathbf{d}_{i+1}$ from Equation 2 are equal to:

$$\mathbf{d}_{i+1} = \mathbf{d}_i + \Delta t \mathbf{v}_i + (\Delta t)^2 (0.5 - \beta)\mathbf{a}_i + (\Delta t)^2 \beta \overline{\mathbf{M}}^{-1}[(1 + \alpha)\mathbf{P}_{i+1} - \alpha \mathbf{P}_i \\ - \mathbf{C}\mathbf{v}_i - (1 + \alpha)(1 - \gamma)\mathbf{C}\Delta t \mathbf{a}_i - (1 + \alpha)\mathbf{r}_{i+1} + \alpha \mathbf{r}_i] \tag{4}$$

where $\overline{\mathbf{M}} = (\mathbf{M} + (1 + \alpha)\gamma \mathbf{C}\Delta t)$. To calculate the displacements $\mathbf{d}_{i+1}$ at the next time step $(i + 1)$ using Equation 4, information from the current time step $(\mathbf{d}_i, \mathbf{v}_i, \mathbf{a}_i,$ and $\mathbf{P}_i)$ and the next time step $(\mathbf{r}_{i+1}$ and $\mathbf{P}_{i+1})$ are required. The externally applied loads $\mathbf{P}_{i+1}$ are known. However, restoring forces $\mathbf{r}_{i+1}$ depend on reaching the displacements $\mathbf{d}_{i+1}$, and therefore Equation 4 is implicit. The displacements $\mathbf{d}_{i+1}$ at the next time step can be written in terms of the predictor displacements $\hat{\mathbf{d}}_i$ (representing the explicit terms in Equation 4) and the remaining implicit terms in Equation 4, whereby:

$$\mathbf{d}_{i+1} = \hat{\mathbf{d}}_{i+1} - (\Delta t)^2 \beta \overline{\mathbf{M}}^{-1} (1 + \alpha)\mathbf{r}_{i+1} \tag{5}$$

Using the equations derived above, Shing et al. (2002) developed an integration scheme for real-time pseudo-dynamic testing by introducing an iterative solution method, where first the predictor displacement $\hat{\mathbf{d}}_{i+1}$ for the next command displacement is computed. The algorithm then proceeds with a fixed number of iteration cycles. Using Equation 5, the command displacements $\mathbf{d}_{i+1}$ for the next step are computed by the use of the restoring forces in each iteration cycle. To have a more or less uniform incremental correction in each iteration step, a scheme involving a fixed number of iterations is used where the computed target displacement $\mathbf{d}_{i+1}^{c(k+1)}$ for iteration k is:

$$\mathbf{d}_{i+1}^{c(k+1)} = \mathbf{d}_{i+1}^{c(k)} + \frac{(\mathbf{d}_{i+1} - \mathbf{d}_{i+1}^{c(k)})}{n - k} \tag{6}$$

In Equation 6, n is the total number of iterations, k is the iteration index, and c is for calculated target command displacement quantities. In the second term of Equation 6, $(n - k)$ is in the denominator, where n is fixed and k increases as the iteration proceeds. That leads to a more or less uniform incremental correction for the command displacements $\mathbf{d}_{i+1}^{c(k+1)}$. A convergence error after the $(n - 1)$th iteration is defined as

$$\mathbf{e}_{i+1}^{R(n-2)} = \mathbf{d}_{i+1}^{m(n-2)} - \hat{\mathbf{d}}_{i+1} + (\Delta t)^2 \beta (1 + \alpha)\overline{\mathbf{M}}^{-1} \mathbf{r}_{i+1}^{m(n-2)} \tag{7}$$

which is used to correct for errors in the displacements $\mathbf{d}_{i+1}$ based on equilibrium and information at the $(n - 2)$th iteration step. By means of these equilibrium corrections performed to eliminate convergence errors, the displacement and restoring force values are made available for the calculation of the predictor displacement $\hat{\mathbf{d}}_{i+1}$ for the next time step while the actuators are imposing the displacement during the last iteration substep. As a result, the structure continues to be loaded in real time without any pause.

The above procedure is similar to one developed by Shing et al. (2002), except that the computed target displacements $\mathbf{d}_{i+1}^{c(k)}$ are used in lieu of the measured displacement of the actuator in the second term of Equation 6. Mercan (2007) determined that this modification improved the accuracy of the testing algorithm. In Figure 3, a multiplication factor λ is also introduced, and will be discussed later.

2.3 *MDOF Real-time hybrid PSD test setup*

The real-time hybrid PSD tests presented in this chapter were performed on the 3-story MRF MDOF structure shown in Figure 4(a). The structure was divided into a 3-story MRF analytical substructure and an experimental substructure consisting of a pair of elastomeric dampers. The MDOF analytical substructure was idealized as a shear building. The story shear-drift hysteretic relationship is shown in Figure 4(b). The story stiffness k, yield dislacement Δy, and strain hardening ρ associated with the MRF floor hysteretic behavior are given in Table 1. The damping matrix for the MDOF system was based on using Rayleigh proportional damping, with a damping ratio of 0.02 in modes 1 and 3. The integration parameters were based on the value of α being set equal to -0.0833.

The analytical substructure was configured as a SIMULINK model, compiled, and loaded onto the Real-Time Target Workstation (see Figure 2) to ensure real-time execution. Using this simplified model, the state determination to evaluate the restoring force developed in the analytical substructure was able to be performed within the specified time interval of the controller's clock speed of $1/1024$ sec, thereby avoiding any delay in the analytical substructure restoring forces. As note previously, any delays in the restoring forces during a real-time hybrid PSD test from either the analytical or experimental substructures can have detrimental effects on the test result. The effect of delay will be discussed later.

The elastomeric damper were comprised of three steel tubes, each tube containing a Butyl blend of rubber that is placed around a longitudinal steel bar and compressed inside the steel tube (see Figure 5(a)), and held in place by friction. The three tubes with the compressed elastomer are placed side by side, as shown in Figure 5(b), and welded together by transverse bars. By attaching the transverse bars between structural members (diagonal braces and a beam), the elastomeric material undergoes shear deformation inside the tube when the frame is laterally loaded. For this study, the dampers are assumed to be attached to both sides of the beam web as shown in Figure 5(c) (the

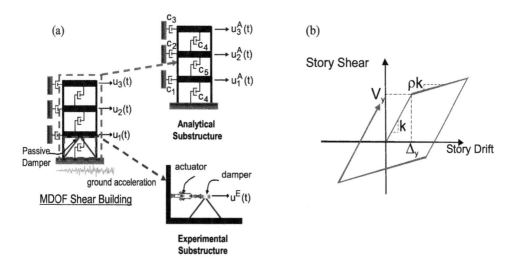

Figure 4.   MDOF real-time hybrid PSD test: (a) Analytical and experimental substructures, and (b) Analytical substructure story shear-drift relationship.

diagonal bracing is not shown in Figure 5(c)). The experiment setup with the hydraulic actuator and the dampers along with a close-up view of the dampers are shown in Figure 5(d) and Figure 5(e), respectively. The instrumentation for the test setup included the actuator load cell, actuator displacement transducer and accelerometer on the actuator clevis. All of the mass of the moving fixtures was weighed; the inertial force developed in these fixtures was found to be negligible.

Material characterization tests on the damper were performed by Kontopanos (2006). Damper properties (stiffness K' and loss factor η) were found to be a dependent on the excitation frequency and deformation amplitude. The damper has the characteristics of an elastomeric material at small deformation amplitudes (of less than 15 mm), with friction dominating the behavior of the damper at larger amplitudes.

2.4 *Effects of time delay in restoring force and velocity feed forward compensation*

During a real-time hybrid PSD test, the restoring force feedback to the integration algorithm may contain a time delay error. The restoring force error from the experimental substructure is due to an actuator control error that does not enable the actuator to achieve the command displacement within a specified interval of time. When a complicated model is employed to represent the analytical substructure, the time taken by the state determination process can cause a time delay in the analytical restoring force.

A restoring force delay introduces additional energy into the system, which may cause the test to go unstable (Horiuchi et al., 1996, Mercan & Ricles, 2007). Mercan (2007) assumed a fixed time delay and used the pseudodelay technique (Gu et al., 2003) to investigate the stability of a time delay in a real-time hybrid PSD test. For a linear time invariant system with time delay, the system dynamics can be expressed in the following state variable form:

$$\dot{\mathbf{x}}(t) = \mathbf{A_{nd}}\mathbf{x}(t) + \mathbf{A_d}\mathbf{x}(t - \tau) + \mathbf{A_{dd}}\dot{\mathbf{x}}(t - \tau) \qquad (8)$$

where $\mathbf{x}(t)$, $\dot{\mathbf{x}}(t)$, $\mathbf{x}(t - \tau)$, $\dot{\mathbf{x}}(t - \tau)$ are the $n \times 1$ state, state derivative, delayed state and delayed state derivative vectors, respectively, associated with n states, and $\mathbf{A_{nd}}, \mathbf{A_d}, \mathbf{A_{dd}}$ are $n \times n$ matrices associated with no delay, delay in the state, and delay in the derivative of the state vector x, respectively. For application to PSD testing, $\mathbf{A_{dd}} = 0$, and the system is known as a retarded delay system. Performing the Laplace transformation of Equation 8, the characteristic equation,

Table 1. MDOF system structural properties.

Floor	Story stiff. - k (kN/m)	Floor mass. - m (mtons)	Yield disp. - Δy (mm)	Strain hardening - ρ
1	11760	135.5	10	0.015
2	11760	135.5	10	0.015
3	9800	67.8	10	0.015

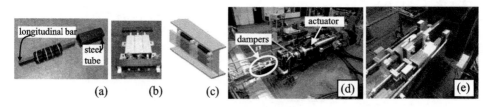

Figure 5. Views of (a) Elastomeric material wrapped around longitudinal bar, and steel tube; (b) Assembled damper; (c) Installed damper; (d) Test setup; (e) Close-up of dampers.

$c(s, e^{-s\tau})$, of the retarded delay system is obtained, where:

$$c(s, e^{-s\tau}) = \det(s\mathbf{I} - \mathbf{A_{nd}} - \mathbf{A_d}e^{-s\tau}) = 0 \qquad (9)$$

By applying an exact mapping of the exponential term, the pseudodelay technique transforms $c(s, e^{-s\tau})$ into a regular polynomial and thereby enables the use of standard methods such as Routh's array to investigate system stability. The pseudodelay technique enables the delay dependent stability analysis of real-time hybrid testing of MDOF systems to be performed where multiple sources of delay exist. The details of the pseudodelay technique and its application to SDOF and MDOF systems are presented in Mercan (2007) and Mercan and Ricles (2007).

By applying the pseudodelay technique to the MDOF hybrid test setup described above, the delay plane shown in Figure 6 is developed. In Figure 6, the normalized analytical substructure delay time $\tau_1(\delta t)$ is plotted on the horizontal axis against the normalized experimental substructure delay time $\tau_2(\delta t)$ on the vertical axis. The delay times are normalized with respect to the digital controller's clock speed of $\delta t = 1/1024$ sec. As can be seen from Figure 6, when there is no delay from the experimental substructure, the system is stable for a delay from the analytical substructure of less than 7 δt (i.e., 0.0068 sec). When there is no delay from the analytical substructure, the system is stable for an experimental substructure delay of up to 85 δt (i.e., 0.083 sec). As noted previously, owing to the RTMD integrated control architecture, the delay from the analytical substructure was eliminated during the real-time hybrid PSD test implementation in this study. To improve the accuracy and stability of the test result, the delay from the experimental substructure was minimized by using a velocity feed forward component in conjunction with a PID feedback controller. The contribution of the velocity feed forward component in reducing the actuator delay will be discussed next.

The frequency response characteristics of a linear system can be investigated using a Bode Diagram. Figure 7 shows the Bode magnitude and phase plots of the combined servo-hydraulic system with a pair of elastomeric dampers for the real-time hybrid test setup. Three different sets of feed forward gains (K_{ff}) for the servo-hydraulic controller were considered. These included: (1) a PID controller with no feed forward ($K_{\text{ff}} = 0$); (2) PID with $K_{\text{ff}} = 2$; and (3) PID with $K_{\text{ff}} = 3$. In all three cases the PID controller gains were set at $P = 20, I = 4, D = 0$. Figure 7 was generated using a linearized servo-hydraulic model and elastomeric damper properties, and provided guidance in the selection of velocity feed forward gain K_{ff} for the real-time hybrid test. From the Bode phase plots in Figure 7, it can be seen that the servo-control loop with only PID control introduces approximately a 23 degree phase lag to an input sinusoid with a 10 rad/sec. frequency. When a velocity feed forward component with $K_{\text{ff}} = 2$ is introduced to the PID controller, the output of the servo-control loop has a small phase lead of approximately 3 degrees with respect to the input. If the velocity feed forward gain is increased to $K_{\text{ff}} = 3$, the phase lead increases to approximately 14 degrees. For all three of these cases, the magnitude Bode plot is around 0 dB at the input frequency of

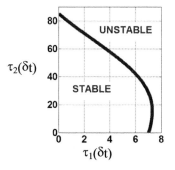

Figure 6.   Normalized stability plane for MDOF real-time hybrid PSD test.

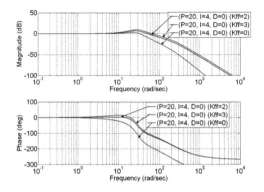

Figure 7.   Bode diagram for combined elastomeric damper and servo-hydraulic system.

10 rad/sec. There is almost no amplitude modification due to the servo-control loop dynamics in the frequency range of 0 to 10 rad/sec. The three natural frequencies of the combined MRF-damper structure are $\omega_1 = 5$ rad/sec, $\omega_2 = 12.8$ rad/sec, and $\omega_3 = 16.9$ rad/sec. The first and second modes of the structure are expected to contribute to the response, with minimal contribution from the third mode. Hence, the PID controller with the feed forward gain of $K_{ff} = 2$ was selected for the real-time hybrid test.

2.5   Real-time hybrid PSD test results

A numerical simulation of the structure without the dampers (i.e., hybrid test without the test structure) was initially performed. The N196E component of the 1994 Northridge earthquake ground motion recorded at Canoga Park, scaled to the Design Basis Earthquake (DBE) was used. The DBE has a 10% probability of being exceeded in 50 years (FEMA, 2003). The first floor lateral displacement time history is shown in Figure 8(a), where it is identified as *undamped frame*. The structure developed yielding, where the first floor displacement exceeded the yield displacement Δ_y of 10 mm, and had a maximum magnitude of 55 mm. As a result of inelastic deformations in the structure, the frame developed a permanent drift at the end of the earthquake, where the first floor static displacement was about 30 mm. The test results for the frame with the elastomeric dampers, where a feed forward control gain of K_{ff} of 2 was used, is also given in Figure 8(a), where they are identified as *damped frame—hybrid test with feed forward*. The hybrid test with the dampers in the test structure resulted in a maximum first floor displacement of 32 mm, where yielding occurred in the first story. The force-deformation of the pair of dampers is shown in Figure 8(b). Energy dissipation is seen to occur, where at a smaller deformation the damper response resembled that of a visco-elastic damper (i.e., the hysteresis loops were elliptical shaped), while at larger deformations (beyond a magnitude of 15 mm) slip occurred and additional energy was dissipated by friction. The slip was not significant, as the dampers nearly self-centered, and a permanent drift of 3 mm occurred after the earthquake.

In real-time hybrid testing the quality of the results, assuming that the analytical substructure has no latency, is best judged by examining the tracking of the hydraulic actuator (Mercan, 2007). A tracking indicator was developed by Mercan (2007) for this purpose. The tracking indicator, TI, is based on the enclosed area of the hysteresis in the synchronized subspace plot, where the actuator command displacement d^c is plotted against the measured actuator displacement d^m (Wallace et al., 2005). The TI is computed at each time step during the test using Equation 10 given below, where at time step i:

$$TI_i = 0.5 \, (A_i - TA_i) \tag{10}$$

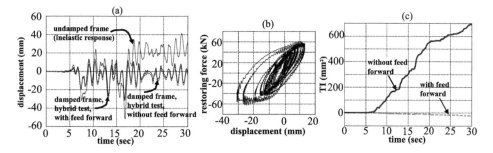

Figure 8.   MDOF Real-time Hybrid Test Results: (a) First Floor Displacement Time History, (b) Damper Hysteretic Response, and (c) Tracking Indicator TI.

In Equation 10, A_i and TA_i are equal to the enclosed area and complementary enclosed area, respectively, where at the beginning of the test, the initial values for A_i and TA_i are set equal to zero, and at time step i:

$$A_i = A_{i-1} + 0.5(d^c_{i+1} + d^c_i)(d^m_{i+1} - d^m_i) \qquad (11a)$$

$$TA_i = TA_{i-1} + 0.5(d^m_{i+1} + d^m_i)(d^c_{i+1} - d^c_i) \qquad (11b)$$

A positive rate of change in the TI implies that the measured displacement is lagging behind the command displacement, whereby energy is introduced into the system, resulting in inaccuracies and possible instabilities in the test. A negative rate of change implies that the measured displacement is leading the command displacement, which adds damping into the system and inaccuracies in the test result. A zero value for the TI implies the absence of a time error during the test. To illustrate the TI, the real-time hybrid test was repeated using the same PID gain settings as before, but without a feed forward component. The results are included in Figure 8(a), where they are identified as *damped frame—hybrid test without feed forward*. The TI's for the real-time hybrid test results presented in Figure 8(a) are shown in Figure 8(c) over the duration of the test. The TI for the test without a feed forward component shows a positive rate of change over time, which indicates that there is an actuator delay throughout the test. The use of a feed forward component with a gain of $K_{ff} = 2$ shows an exceptional tracking capability by the servo-controller, where the slope for the corresponding TI is orders of magnitude smaller than that without a feed forward component. The measured actuator displacements without a feed forward component has a delay of 20.5 δt (0.02 sec.) behind the command displacement, and with a feed forward component a time lead of about 0.5 δt (0.0005 sec.). On the basis of the values for the TI, the results from the hybrid test with a feed forward component are considered to be exceptional, and more accurate than the results from the test without a feed forward component.

2.6   *Determination of damper requirements to meet design criteria*

A performance-based design procedure involving the use of elastomeric dampers in steel frame systems was developed by Lee et al. (2005). The procedure involves selecting seismic performance objectives associated with a seismic hazard level, and performing the damper design by considering a range of values for the damper stiffness relative to the frame lateral story stiffness and brace stiffness, as well as the structural characteristics of the frame (natural period, first mode shape, member forces resulting from seismic loading). Lee et al. (2006) show that it is possible to economically design a steel frame with elastomeric dampers, where the maximum drift is less than 1.5% and the members in the steel frame remain essentially elastic under the DBE ground motions. For the elastomeric dampers considered in this chapter, the combined frictional and elastomeric behavior

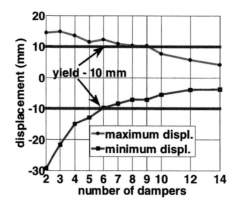

Figure 9.   Analytical substructure first story maximum and minimum displacements as a function of the number of dampers.

of the damper make it difficult to assign a value for K' and η for use in the design procedure, since these properties are dependent upon both frequency and deformation amplitude. If an accurate analytical model does not exist for the damper, an assessment of whether the dampers enable the performance objectives of the structure to be met based on time history analysis would not be reliable.

The real-time hybrid testing method is ideal to evaluate structural performance under dynamic loading when the behavior of components of a structural system cannot be accurately modeled. Consequently, the test method was used to establish the damper design requirements (total number of elastomeric dampers required) to satisfy the performance objective of having the first floor of the 3-story frame remain elastic under DBE level ground motions. To accomplish this, the existing test setup described above was used. The real-time hybrid testing algorithm was modified, where the measured restoring force r_{i+1}^{m} from the experimental substructure was multiplied by a factor λ in order to represent a specific number of dampers being simulated in the structural system at the first floor (see Figure 3). By performing a series of real-time hybrid tests, each with a different value for λ, the number of dampers required to have the first floor of the MRF remain elastic could be determined. The existing test setup represented the case where two dampers were used (where $\lambda = 1$).

Eleven real-time hybrid tests were performed, where the structural system was subjected to the same DBE-scaled Canoga Park ground motion and λ was varied from 1 to 7. Figure 9 presents the maximum and minimum first story displacements developed in the structure as a function of the number of dampers. From the test results it was determined that eight dampers (i.e., $\lambda = 4$) are required to keep the first floor displacement within the elastic range (i.e., less than 10 mm.).

3   CONCLUSIONS

The real-time hybrid PSD test method offers a means to investigate the dynamic performance of complex structures by combining computer simulation with physical testing. The coordination of the experimental and analytical substructures is critical, where in order to ensure stability and accuracy, during a hybrid test the commands and feedbacks from both substructures need to be synchronized. The time delay from the analytical substructure restoring force due to the time taken by the state determination computations, transmission and processing of signals should be minimized. Through the use of Lehigh RTMD integrated control architecture, the analytical substructure restoring force delay was eliminated in this study. Similarly, the time delay from the experimental substructure restoring force due to servo-hydraulic system dynamics needs to be minimized. In this study, the use of a velocity feed forward component combined with a PID feedback controller enabled the tracking of the command displacements by the hydraulic actuator to be improved, significantly reducing the time delay introduced by the experimental substructure. In order to assess the quality of the test results, a tracking indicator was proposed to enable the tracking of the actuator during a test. The real-time hybrid test method enabled a determination of the effectiveness of complex elastomeric dampers, which are sensitive to both loading rate and deformation amplitude, to reduce potential seismic damage to an MRF under the DBE level earthquake.

ACKNOWLEDGEMENTS

Funding for this research was provided by the National Science Foundation under Grant No. CMS-0402490 within the George E. Brown, Jr. Network for Earthquake Engineering Simulation Consortium Operation, and by the Pennsylvania Department of Community and Economic Development. The authors wish to thank Corry Rubber Company for donating the elastomeric dampers used in this research. In addition, the authors would like to express their appreciation to Thomas Marullo of ATLSS Research Center for his contributions during the experimental studies reported here.

REFERENCES

Dermitzakis, S.N. and Mahin, S.A. (1985) "Development of substructuring techniques for on-line computer controlled seismic performance testing", *Report UBC/EERC-85/04*, Earthquake Engineering Research Center, University of Cal, Berkeley.

FEMA (2003) "NEHRP Recommended Provisions for New Buildings and Other Structures, Part 1—Provisions," Report No. FEMA 450, Federal Emergency Management Agency, Washington, D.C.

Gu, K., Kharitonov, V.L. and Chen, J. (2003) *Stability of Time-delay Systems*, Birkhäuser, Boston, MA.

Hilber, H.M., Hughes, T.J.R. and Taylor, R.L. (1977) "Improved numerical dissipation for time integration algorithms in structural dynamics," *Earthquake Engineering and Structural Dynamics*, 5, 283–292.

Horiuchi, T., Nakagawa, M., Sugano, M. and Konno, T. (1996) "Development of a real-time hybrid experimental system with actuator delay compensation," *Proc of the Eleventh World Conference on Earthquake Engineering*, Paper No. 660.

Kontopanos, A., (2006) "Experimental investigation of a prototype elastomeric structural damper," *MS. Thesis*, Lehigh University, Bethlehem, PA.

Lee, K., Ricles, J. and Sause, R. (2006) "Performance based seismic design of steel MRFs with elastomeric dampers," *Journal of Structural Engineering*, submitted for publication.

Lee, K.-S., Fan, C.-P., Sause, R., and Ricles, J., (2005) "Simplified design procedure for frame buildings with viscoelastic or elastomeric dampers," *Journal of Earthquake Engineering and Structural Dynamics*, 34, 1271–1284.

Mercan, O. and Ricles, J.M. (2007) "Stability and accuracy analysis of outer loop dynamics in real-time pseudo-dynamic testing of SDOF systems," *Earthquake Engineering and Structural Dynamics*, in press.

Mercan, O., (2007) "Analytical and experimental studies on large-scale, real-time pseudodynamic testing," *Ph.D. Dissertation*, Lehigh University, Bethlehem, PA.

Shing, P.B., Spacone, E., Stauffer, E. (2002) "Conceptual design of a fast hybrid test system at the University of Colorado," *Proc 7th U.S. National Conference EQ Eng*, Boston, MA.

Simulink (2007) The Math Works, Inc., Natick, Massachusetts.

Wallace, M.I., Sieber, J., Nield, S.A., Wagg, D.J. and Krauskopf, B. (2005) "Stability analysis of real-time dynamic substructuring using delay differential equations," *Earthquake Engineering and Structural Dynamics*, 34, 1817–1832.

CHAPTER 11

CU-NEES fast hybrid testing facility

G.J. Haussmann, V.E. Saouma & E. Stauffer
University of Colorado, Boulder, USA

ABSTRACT: Following a brief discussion on the merits of fast hybrid testing, this paper first describes the CU-NEES facility in terms of hardware and software. Current development (such as the desktop FHT system), and recent completed projects are described next. Finally, current research and development is highlighted.

1 INTRODUCTION

Our laboratory is a member of the Network for Earthquake Engineering Simulation (NEES) program funded by the National Science Foundation (NSF). As one of the smallest testing facility, our emphasis has been on software development toward achieving real time hybrid simulation of meaningful models.

Fast Hybrid Testing (FHT), a cost effective alternative to expensive and complex shake table tests, enables us to zoom into the rate dependent substructure which may be problematic and test it physically (possibly full scale) while a finite element (typically) analysis simulates its complementary substructure which is "well understood". There are many advantages to this modern testing paradigm: 1) Avoidance of "background noise" stemming from a complex (and often scaled down) shake table test if we are primarily interested in a particular detail/component (facilitates component design verification); 2) Ability to perform multiple simulations with minimal structural damage (specimens can be relatively inexpensively replaced, or for energy dissipators no replacement is necessary); 3) Possibility of simulating different structures with the same physical substructure (for instance the same physically tested energy dissipator can be part of different structural arrangements); and 4) fully coupled analysis. Finally, if shake table tests constitute a full physical simulation, and an ultimate objective is to achieve full numerical simulation

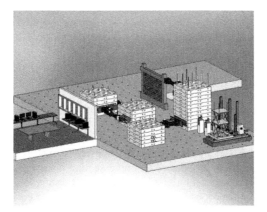

Figure 1. CU-NEES laboratory facility.

(as computational fluid dynamics has progressively replaced wind tunnel tests), then hybrid simulation is a required and indispensable step toward this ultimate goal.

2   MOTIVATION

The emphasis at the CU NEES site is on both *Fast* and *Realtime* hybrid testing, where these terms are defined as:

Realtime tests are performed at full speed, so that a hybrid test meant to simulate thirty seconds of building behavior will take exactly thirty seconds to perform. This is contrasted to many hybrid tests where the thirty seconds of simulation may take several hours or days to perform in the lab as calculations are done and actuators are re-positioned.

Fast testing is a test performed at a slow speed, slower than realtime. However, the ratio of simulation time to actual time is constant. For example, a test performed at one-half speed would be consider a Fast Hybrid Test.

We also refine the concept of realtime into "hard" and "soft" kinds as (Stankovic, 1998): *Hard Realtime* tests are realtime tests where the timing constraints are strictly met. For instance, a hard realtime simulation with an update rate of 1,000 Hz, would require a new data update every 1 mS. All computations and data transfer for the update must be completed within a 1 mS; if the update takes longer than 1 mS due to *nondeterministic* processes, then the entire hard realtime test is considered invalid.

Soft Realtime tests are realtime tests where the timings requirements are somewhat relaxed. As an example instance, if a soft realtime test was performed with an update rate of 1,000 Hz (1 mS), occasional updates that took longer (1.5 mS, 2 mS) are still considered valid updates.

Deterministic processes are processes such as computer programs and data transfers that are guaranteed to complete in a certain amount of time. Most modern computer software and hardware technology—such as the Microsoft Windows Operating System, and the ethernet networking protocol—were not built as deterministic systems to reduce cost and development time.

The technical requirements for fast and hard realtime hybrid tests imply strong constraints on the possible tests that can performed. In order to keep the physical specimen moving, new data commands must be sent to it at a relatively fast rate. As an example, the MTS controller at the CU NEES site sends updates 1,024 times a second, or every 0.97 milliseconds; herefore the computer model of the structure must also provide new data every 0.97 milliseconds, incorporating measurement information from the previous update. In fact, every timestep of the computer simulation has to complete in 0.97 milliseconds, which adds a *Hard Realtime* constraint. On most modern computers running modern operating systems such as Microsoft Windows or Apple MacOS, programs "hiccup" or "freeze" for a fraction of a second while the computer tends to a more important task such as disk access, memory cleanup, etc. If one of these "hiccups" occurred during a hybrid test, then the result would be disastrous—updates would not be sent to the controller hooked up to the physical specimen, resulting in an incorrect and/or destructive response from the specimen.

In addition, the integration method is constrained. The short time allowed for each update imposes a limit on the complexity allowed for the integration method, so that higher-order integration methods such as the Dormand-Prince integrator—a fourth-order Runge-Kutta style method used in Simulink—are unattractive choices. The fixed update rate also constrains the timestep size of the simulation model to a specific value, which can pose a problem for explicit integration methods. Explicit methods can be unstable with large timesteps, and the stability problem cannot be solved by simply making the timestep smaller, because the timestep is fixed.

The Hard Realtime constraint is satisfied by implementing the computer model on a Realtime Operating System, where computation can be guaranteed to complete by a specific deadline. By combining these design decisions with an implicit timestepping method the CU NEES site can operate a real-time hybrid test with multiple nonlinear elements. However, it should be noted that these requirements will limit the size and complexity of the numerical model.

3   LABORATORY DESCRIPTION

3.1   *Physical facility*

The structural lab hosting CU NEES is a 70 by 30 ft strong floor with a 3′ grid anchoring matrix (accesible from both sides). Twenty six modular concrete square blocks (88 by 12 in.) can be tied down through Diwydag vertical ties to provide vertical reactions walls.

The facility is equipped with the following high performance hydraulic actuators: 1 MTS 244.51S actuator (220 kip, 12 in. stroke and 10 in/sec. velocity), 2 MTS 244.41S (110 kip, 10 in. stroke and 20 in/sec. velocity), and 4 MTS 244.22 (22 kip, 24 in. stroke and 100 in/sec. velocity). The hydraulic power unit has a maximum flow capacity of 180 GPM.

3.2   *Data communication*

The CU-NEES site operates a network of computers to perform the multiple tasks involved before, during, and after a hybrid simulation, Fig. 3.2. The coordination of these machines is done by two networks: a gigabit-ethernet network for data sharing and bulk data transfer, and a SCRAMnet for realtime communication.

The bulk of user interaction is done via several host computers. On these host computers the user can configure various parameters of the hybrid test, such as modeling the structure to be analyzed, select data quantities to capture, and which actuators to activate during a simulation. This configuration is performed before the hybrid simulation, and files needed for configuration such as ground motions, material parameters, and others are transferred using the ethernet network. Similarly, after a hybrid simulation the collected data is stored in a local repository and possibly sent to a remote site for storage and/or visualaization again using the ethernet network. However, while the ethernet network is used extensively both before and after a hybrid test, the SCRAMnet network is use primarily during a hybrid test.

During a fast hybrid simulation, there are multiple computer platforms running simultaneously, each performing a different task associated with the simulation. One computer (running a real time operating system) will be performing the numeric simulation of the structure, while another will communicate with and control the physical test equipment, and a third computer might be storing data generated by the test. All these computers need to communicate with each other during the simulation in a near instantaneous fashion.

While the standard ethernet-based network would work for a pseudodynamic hybrid test, ethernet as a network protocol is unsuitable for fast and realtime hybrid simulations due to the nondeterminism of the ethernet network. Specifically, ethernet is an example of a "best-effort" network; if computer A tries to send data to computer B over ethernet, there is no upper limit to the transmittal time, (Many, 2005). When results have to be analyzed and produced 1,024 times a second, the unpredictability of ethernet is a serious liability for fast and realtime hybrid tests. To this end, a

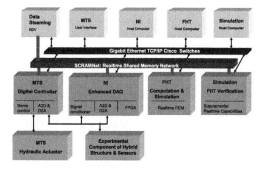

Figure 2.   The overall network architecture of the CU-NEES site.

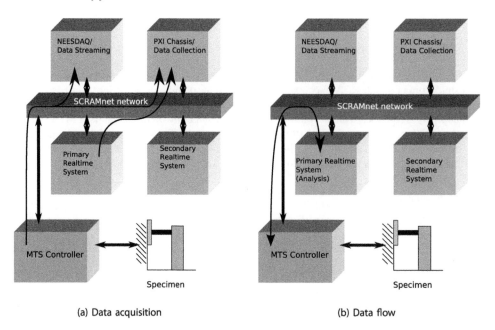

(a) Data acquisition   (b) Data flow

Figure 3.   SCRAMnet card features.

network which has guaranteed delivery time as one of its characteristics is adopted. The SCRAMnet network is a deterministic data network that appears as distributed shared memory to all the computers using it. When a computer on the SCRAMnet writes data to a particular memory location in the SCRAMnet memory space, that data appears on the memory space of all the other computers, connected to the SCRAMnet. This effect is deterministic; the data transfer is guaranteed to occur within a specified time, on the order of nano-or micro-seconds. The end result is that data results provided by the simulation computer are guaranteed to be sent to the actuator controller in a timely manner. Using a different network—such as ethernet—it would be possible that data could be delayed or dropped altogether, such that the actuator controller will not have any new data for the next 0.97 millisecond update.

The numeric computation portion of a fast hybrid test runs on the primary realtime host, which is a standard PC with a realtime operating system (Phar-Lap ETS) and simulation software, Fig. 3. This PC is connected via SCRAMnet to the actuator controller and the data collection platform. Although the computation software currently used is a heavily customized version of OpenSEES, (Wei, 2005) any simulation software could be used if properly modified to run on the realtime operating system.

Various other tasks related to a hybrid test can be performed on the secondary realtime host, which is also a standard PC with a realtime operating system and SCRAMnet hardware. The secondary realtime host serves as a backup of the primary realtime host, but can perform other tasks as well, such as simulating the response of actual test hardware or processing data as it flows through the SCRAMnet network.

Data is gathered by three separate platforms, for different purposes, Fig. 3. The first platform is the management computer directly connected to the MTS controller. This management PC allows the viewing and saving of various control quantities for diagnostic purposes. For instance, all the data being measured by instruments in the MTS controller are available to the management computer, (Wallen, 2007).

In addition, the MTS controller writes all of its control and measurement quantities to specific memory locations in the SCRAMnet memory space, so that other computers on the SCRAMnet can

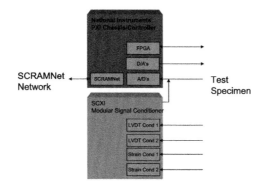

Figure 4.   Hybrid data acquisition system.

view, process, and possibly record these values. The second platform, a data streaming computer platform, reads all these data values from the SCRAMnet and pushes them onto the CU-NEES data turbine, which allows for remote viewing of the data quantities using RDV, (*R.D.V. User's Guide* 2007). This is used to view realtime, streaming data over the network for a data turbine server, synchronizes numeric, video and still images, and allows for instant replay with variable speed playback.

The third platform is a National Instruments PXI chassis containing a PC, SCRAMnet hardware, and a panoply of other instrumentation I/O, Fig. 3(a) for (locally written) Hybrid Data Acquisition (DAQ) system. It plays a key role in the Fast Hybrid Testing (FHT) system by acquiring time synchronized data from test specimen mounted physical sensors and node/element data from within the numerical model in real time, Fig. 3.2. Physical sensor data is acquired using conventional National Instruments multi-function IO hardware and SCXI signal conditioning hardware, while numerical model data is acquired using a SCRAMNet network interface card. The SCRAMNet network is the communication backbone of the FHT system. All feedback and command signals are exchanged at a rate of 1024 Hz across the SCRAMNet. Using a LabVIEW Realtime program, the Hybrid DAQ acquires a preprogrammed list of SCRAMNet channels using the MTS generated network interrupt as a clock signal. The system can acquire over 500 channels at 1024 Hz. in its current configuration. All acquired data is written to a Matlab format file or a tab delimited text file.

3.3   *Software*

To compensate for its relatively limited physical capability CU NEES site has placed much emphasis on software capability. Currently, numerical simulation is performed by OpenSEES which was modified to run completely on a realtime operating system, with direct access to the actuator controller, (Wei, 2005). More recently, Haussmann (2006c) "dissected" the code, and wrote a separate software-independent, library to connect a finite element code with the hardware.

As the software simulation runs, computation results from the computations are written to the SCRAMnet network, which are then used by the MTS controller as a new position command for the actuators, Fig. 3(b). Then, in turn, measurements are collected by the MTS controller and written back to the SCRAMnet, which are then picked up by the software simulation as a restoring force to be used in subsequent computations. This entire sequence occurs every 0.97 milliseconds, or 1,024 times per second. Recent benchmarking indicated that we can support approximately 135 degrees of freedom in hard real time simulation of a nonlinear frame.

3.4   *Five levels of simulation*

Experience at the CU NEES facility has shown a relatively simple five level approach to hybrid simulation to be effective.

Level one is a pure simulation with a 100 percent numerical model. Level two divides the structure into a numerical component and experimental component which is again simulated fully on a single computer with numerical modeling (FEM). Level three places the numerical model on one computer and a typically linear representation of the experimental component on a second computer. Level four replaces the linear representation of the experimental component with the real test component and is an actual fully integrated hybrid simulation. A level five would be used if an external controller is implemented.

Hence prior to conducting a hybrid simulation, we would start with the first level, and gradually move up to the fourth, and possibly to the fifth if an external controller is used.

3.5   *Integration scheme*

The hybrid simulation capabilities at the CU NEES facility are based on a constrained implementation of the α method, (Hilber, Hughes and Taylor, 1977). The favorable stability and damping properties of this method and its successful application to conventional pseudodynamic tests (Shing, Vannan and Carter, 1991) make it well suited for a fast and continuous testing system. With the understanding that the externally applied force vector is balanced by inertial, damping, and restorative force components and that each of these components is composed of contributions from both the numerical (FEM) and experimental portions of the hybrid simulation the damping term is generalized so as to admit nonlinear behavior.

$$Ma + s(v) + r(x) = f \tag{1}$$

When considering a hybrid structure each of the three terms $Ma, s(v)$ and $r(x)$ may be expanded so as to make explicit the hybrid nature of this representation.

$$Ma = (M_{exp} + M_{FEM})a \tag{2}$$

$$s(v) = s_{exp}(v) + C_{FEM}(v) \tag{3}$$

$$r(x) = r_{exp}(x) + r_{FEM}(x) \tag{4}$$

In a traditional pseudo-dynamic test the first term on the right hand side of Eq. 3 and 4 is neglected owing to the relatively small magnitude of these components.

Indeed, when a hybrid test distorts the representation of time by slowing and often stopping the passage of simulation time these terms are negligible or nonexistent. For some materials and devices which might be tested in a hybrid simulation these conditions may be tolerable. On the other hand there are materials and devices, such as an MR Damper, that are rate or velocity sensitive and these terms must be properly accounted for.

The discrete time equilibrium equations for this representation of a second order dynamic system are

$$Ma_{i+1} + (1 + \alpha)s_{i+1} - \alpha s_i + (1 + \alpha)r_{i+1} - \alpha r_i = (1 + \alpha)f_{i+1} - \alpha f_i \tag{5}$$

$$d_{i+1} = d_i + \Delta t v_i + \Delta t^2 \left[\left(\frac{1}{2} - \beta \right) a_i + \beta a_{i+1} \right] \tag{6}$$

$$v_{i+1} = v_i + \Delta t \left[(1 - \gamma) a_i + \gamma a_{i+1} \right] \tag{7}$$

Where M is the mass matrix, s is the damping force vector assumed to be a nonlinear function of the velocity vector, r and f are the restoring force and external applied force vectors respectively.

This direct method of time integration determines equilibrium at equally spaced time intervals which herein will be referred to as the integration interval. In order to allow for nonlinear structural response it necessary to include an iteration capability that converges to the equilibrium condition within each integration interval. A modified Newton-Raphson iteration method is applied to the discrete equations of motion (equation 1). The finite number of iterations, which will be constrained to a fixed and constant number, (Shing, Spacone and Stauffer, 2002), act to subdivide the integration interval into n iteration intervals. If each interval is given equal time weighting the iteration interval may be expressed as

$$\delta t = \Delta t / l \tag{8}$$

Where l is the integer number of Newton iteration intervals, Δt and δt are the time intervals associated with the integration interval and iteration intervals respectively. By fixing l to be a constant integer value a favorable degree of determinism is achieved which is important for realtime integration and hybrid testing. This determinism comes at the price of constraining the calculation of equilibrium to a limited number of Newton iterations and a fixed interval of time in the case of consistently scaled and realtime simulations. Experience at the CU NEES FHT facility has indicated that $\Delta t = 0.01$ and $\delta t = 0.001$ are reasonable values that balance the need for accuracy and speed.

As Fig. 5 illustrates, a force correction is need at each iteration. The force correction accounts for both actuator position error and the interpolated displacement value. The linear initial stiffness K_i is used to project the force forward to a value consistent with the target displacement d_{n+1}^k reducing the effects of one possible source of error

$$r_{n+1} = r_{n+1}^{M(k)} + K_i(d_{n+1}^k - d_{n+1}^{M(k)}) \tag{9}$$

Ideally a similar correction is applied to the velocity dependant measured damping force s_{n+1}^M but is not currently implemented due to the lack of an accurate and reliable velocity measurement device that may be positioned at each of the physical degrees of freedom.

The iterative solution procedure is based on the linearized Taylor series representation of the residual equilibrium equation

$$f_r(d_{n+1} + \Delta d) \approx f_r(d_{n+1}) + \frac{\partial f_r}{\partial d_{n+1}} \Delta d \tag{10}$$

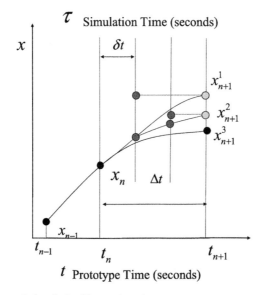

Figure 5.   Command interpolation during Newton iteration.

Where successive displacements increments Δd are computed until a convergence criterion is satisfied. With appropriate consideration of the nonlinear damping and stiffness terms the Jacobian may be expressed in the form

$$\frac{\partial f_r}{\partial d_{n+1}} = M + c_1 \left(\frac{\partial s_{n+1}}{\partial v_{n+1}} \frac{\partial v_{n+1}}{\partial d_{n+1}} + \frac{\partial r_{n+1}}{\partial f_{n+1}} \right) \tag{11}$$

Both the stiffness and the damping terms are approximated with the linearized initial stiffness and initial damping matrices K_i and C_i and a single equation is obtained which is repeatedly solved until the desired level of accuracy is obtained

$$d_{n+1}^{k+1} = d_{n+1}^k - \left[M + c_1 \left(\frac{\gamma}{\Delta t \beta} C_i + K_i \right) \right]^{-1} f_r(d_{n+1}^k) \tag{12}$$

The indices n and k indicate the integer value for the time step and the Newton iteration number respectively. In preliminary testing this new integration scheme has proven to be equal or superior to the prior scheme (restricted to linear viscous damping) both in terms of accuracy and rate of convergence.

4   DESKTOP FHT

While the CU-Boulder NEES site has implemented a complete FHT solution, the high cost and effort involved in setting up and running an FHT test may prohibitive for certain applications, including: 1) Live demonstration of the FHT method in offsite locations; 2) Interactive experiments or classroom presentations; and 3) Fast prototyping of new engineering ideas or concepts, (Haussmann, 2006b). To address these problems, a high performance desktop platform for realtime hybrid simulation is being developed at CU NEES. The hardware and software requirements will provide basic FHT functionality with relatively lower cost and more portability than full-scale hybrid test sites.

The desktop platform includes a full implementation of the fast hybrid method used in the CU production FHT test lab. A multiple-DOF simulation analyzes the structure while the actuators and instruments (available through LabView VI modules) drive the test specimen and inject measurements back into the simulation. The simulation method uses an implicit scheme for stability and an iterative solution method to handle nonlinear responses.

The simulation and instrumentation are handled by a single computer platform, called the target platform. The target platform uses a real-time operating system to insure a deterministic response for the hybrid test. The target platform sends simulation state and other information over a network to a second computer, called the host computer. The host computer runs a constantly updated visualization to show the states of the simulation structure for debugging and demonstration. The host platform also allows for user interaction to start and stop the simulation, and reconfigure the instrumentation as accessed in LabView.

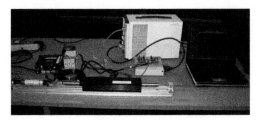

Figure 6.   CU-NEES desktop system.

5 REPRESENTATIVE PROJECTS

5.1 *Zipper Frame*

The Zipper Frame structure is a distributed pseudodynamic hybrid test performed jointly by the University of Colorado at Boulder and the University of California at Berkeley, (Yang, Stojadinovic and Moehle, 2006). The structure involved was a braced steel frame comprised of multiple components. The simulation was a distributed test—that is, the test was not performed entirely at one site, but instead multiple sites coordinated their efforts to jointly perform a hybrid test, Fig. 7.

The Zipper Frame structure was excited with various ground motions. One ground motion was a low-intensity excitation where the materials of the structure remained mostly elastic. Later tests included more intense excitations which produced plastification and buckling in the metal beams.

5.2 *Magnetorheological damper*

A separate chapter in this monograph describes a recently completed NEES-R project *Experimental Verification of Semiactive Control of Nonlinear Structures Using Magneto-Rheological Fluid Dampers* by Prof. Christenson. It involves the use of semiactive components—dampers with a dynamically variable viscosity.

The project centers around the use of Magnetorheological dampers, which can have their damping coefficient controlled by a variable current. By varying the current running through a wire coil embedded in the damper, the viscosity of the damper fluid is changed, increasing or decreasing the damping. With the proper drive circuitry, the drive current can be change from zero amps (minimal/no damping) to 2.5 amps (maximum damping) in the space of 30 mS. This allows the damping in a building to be changed dynamically in response to earthquake events.

Different damper behaviors can be produced by reprogramming the damper control method, without any costly replacement or mechanical modification of the dampers themselves. Tests done

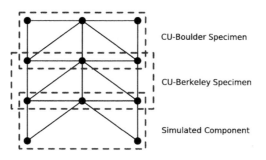

Figure 7. Distributed hybrid simulation.

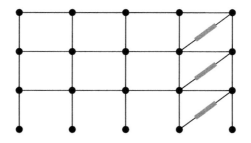

Figure 8. Simulated building used for the MR damper NEES-R project.

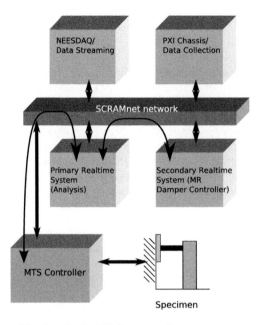

Figure 9.   Data transfer modifications for the MR-damper project.

at the CU-NEES site are intended to examine the use of various control methods to examine their effectiveness.

5.3   *Modification of the hybrid test method*

The unusual nature of the MR dampers required that they receive a control current to vary the amount of damping. In a fast hybrid test, these control current need to be produced in real-time, computed from the forces on the dampers and from specific quantities in the building itself. This requirement leads to some unusual needs that are outside the scope of a typical hybrid test, and resulted in several key modifications and additions to the CU-NEES site testing capabilities.

The first modification required was due to the nonlinear, time-varying nature of the physical specimen. Typical hybrid tests assume the restoring force of the physical specimen is a displacement-based springlike response and structure their computations accordingly. To accommodate the MR damper, additional computation assuming a nonlinear damper response was added to the hybrid software. The fact that the actual damping can vary over a large dynamic range affects the setting of key configuration parameters: when setting the parameter values, we must base the values on the assumed largest possible damping allowed, in order to avoid stability problems caused by large restoring force produced by the MR damper.

The second modification was to allow for the use of a controller, which requires access to accelerations and forces in the software simulation and produces control signals which control the physical damper specimens. To accomodate this requirement, the simulation software was augmented with several additional components which write simulation quantities—such as certain nodal accelerations, and the ground motion—to the SCRAMnet which is then picked up by the controller software. In addition, interpolation and time-shifting must be done to cross the "hybrid gap" that separates the running software simulation and physical specimen; this must be done because the controller receives input from the software side but produces control currents that operate on the hardware side.

6 CURRENT RESEARCH/DEVELOPMENT

Aside from providing "routine" support to the research community, our center maintains an active research and development activity primarily focusing on software and control. Among the currently ongoing efforts:

1. Trajectory exploration approach to dynamic testing, in particular hybrid simulation. The idea is to seek inputs to the dynamic testing device (for example, shaking table)-tested system combination, that will create interesting behavior such as periodic trajectories (our current work), failure modes etc, (Prof. Sivaselvan and Prof. Hauser).
2. Usage of feedforward techniques for dynamic testing using measurement of dual variables for dynamic compensation, (Prof. Sivaselvan and Mr. Jochen).
3. Use Field-Programmable Gate-Array (FPGA) hardware to perform large scale numerical simulations. Preliminary tests, Haussmann (2006a) indicate that we can potentially increase by at least a factor of 100 the number of degrees of freedom at about 40 MHz (as opposed to about 40 at 1 MHz with current capabilities), (Dr. Haussmann).
4. Exploration of the possibility of using a cluster of parallel computers to perform the numerical calculations of the hybrid simulation, (Jeremic, 2007), (Prof. Saouma).
5. Development of a streamlined finite element analysis software written explicitly for hard real-time hybrid simulation and which could eventually accommodate very large structural analysis problems through advanced element and constitutive model formulations, (Prof. Saouma, Dr. Haussmann and Mr. Kang).
6. OpenFresco and UI-SIMCOR are currently being "localized" in order to interoperate with other NEES sites using these software libraries for hybrid testing (Dr. Haussmann).

ACKNOWLEDGMENTS

CU-NEES is funded by the George E. Brown, Jr. Network for Earthquake Engineering Simulation (NEES) program of the National Science Foundation under Cooperative Agreement CMMI-0402490 and from the University of Colorado in Boulder. The authors would also like to thank Prof. B. Shing who was the initial proponent and Director of this facility, Mr. R. Wallen and Mr. Tom Bowen of the CU-NEES site.

REFERENCES

Haussmann, G.: 2006a, Analysis and calibration process for hybrid simulations using fixed-point arithmetic, *Technical Report CU-NEES-06-5*, CU-NEES Fast Hybrid Testing Laboratory.
Haussmann, G.: 2006b, The cu-nees fast hybrid test desktop platform, *Technical Report CU-NEES-06-3*, CU-NEES Fast Hybrid Testing Laboratory.
Haussmann, G.: 2006c, Hybrid simulation hardware interface, *Technical Report CU-NEES-06-7*, CU-NEES Fast Hybrid Testing Laboratory.
Hilber, H., Hughes, T. and Taylor, R.: 1977, Improved numerical dissipation for time integration algorithms in structural dynamics, *Earthquake Engineering and Structral Dynamics* 5, 283–292.
Jeremic, B.: 2007, Preliminary report on using high performance parallel finite element analysis for fast hybrid simulation, *Technical Report CU-NEES-07-15*, CU-NEES Fast Hybrid Testing Laboratory.
Many: 2005, Ieee standard for information technology–telecommunications and information exchange between systems–local and metropolitan area networks– specific requirements–part 3: Carrier sense multiple access with collision detection (csma/cd) access method and physical layer specifications. IEEE Std 802.3-2005, by the IEEE 802.3 Working Group.
R.D.V. User's Guide: 2007. http://it.nees.org/documentation/pdf/rdv-15-users-guide.pdf.
Shing, P.B., Spacone, E. and Stauffer, E.J.: 2002, Conceptual design of fast hybrid test system at the university of colorado, *Proceedings of the Seventh U.S. National Conference on Earthquake Engineering*, Boston.

Shing, P., Vannan, M.T. and Carter, E.: 1991, Implicit time integration for pseudodynamic tests, *Earthquake Engineering and Structural Dynamics* **20**, 551–576.

Stankovic, J.: 1998, *Deadline scheduling for real-time systems: EDF and related algoritms*, Klewer Academic Publishers.

Wallen, R.: 2007, Scramnet labview data acquisition system, *Technical Report CU-NEES-07-16*, CU-NEES Fast Hybrid Testing Laboratory.

Wei, Z.: 2005, *Fast Hybrid Test System for Substructure Evaluation*, PhD thesis, University of Colorado, Boulder.

Yang, T., Stojadinovic, B. and Moehle, J.: 2006, Hybrid simulation evaluation of innovative steel braced framing system, *Proceedings, Eighth National Conference on Earthquake Engineering*, San Francisco. Paper #1415.

CHAPTER 12

Real-time substructure testing on distributed shaking tables in CEA Saclay

J.C. Quéval & A. Le Maoult
Commissariat á l'Energie Atomique (CEA), Saclay, France

U.E. Dorka & V.T. Nguyen
Steel and Composite Department, University of Kassel, Germany

ABSTRACT: The CEA Saclay Laboratory is working on different national and international programs related to the seismic behavior of buildings, structures, components and equipments. With the seismic testing facility, many kinds of tests for model development, validation, calculation methods development, qualification, codification, and qualification of components can be carried out. One powerful and potential testing method that should be developed for the need of CEA Saclay is Real-Time Substructure Testing (RTST). Moreover, an available testing algorithm developed by Dorka et al. can be applied and developed further for RTST by using shaking tables in CEA Saclay. From these, the project of common action between CEA Saclay and the University of Kassel has been set up.

In this project, CEA Saclay has built a model including two storey steel frame with two Tuned Mass Dampers (TMD) and is conducting series of dynamic tests of full model on a shaking table, while the University of Kassel has implemented the algorithm based on the formulation from the result of a European project (the DFG-project Do 360/7), developed a software for RTST and now is going to conduct series of RTSTs of two TMDs on two distributed shaking tables. By carrying out both dynamic tests of full model (reference tests) and RTSTs of TMDs, the algorithm and the results of RTSTs can be verified.

This paper is going to introduce current achievement of the project including (1) the development of control system for distributed substructure tests on shaking tables in CEA Saclay, (2) the tested model including a two storey frame with two TMDs, (3) testing simulation of RTST of TMD and testing plan and (4) the discussion on application of RTST on distributed shaking tables and upcoming work in the project.

1 INTRODUCTION

Seismic testing facility (TAMARIS) in CEA Saclay is powerful for not only caring out dynamic tests but also for investigating on field of substructure tests. TAMARIS has largest test facilities within the European Community and also has the largest shaking table in Western Europe [1]. TAMARIS is composed of 4 shaking tables (VESUVE, TOURNESOL, MIMOSA, AZALEE), a 15 depth m pit (IRIS), a reaction wall and accompanied facilities for measurement and control of testing processes. Over 15 years, these facilities have been developed and used mainly for performing dynamic tests of R&D and seismic qualification in a wide range of applications of buildings, structures, components and equipments. In front of the need of development, TAMARIS should develop further in order to be able to perform sub-structure tests for applications with lager structures in civil engineering.

On the other hand, an available testing algorithm developed by Dorka et al. can be applied and developed further for RTST [2, 3, 4]. Since 1998, a project researching on the algorithm and experimentation of real-time substructure tests in resembling aerospace application was conducted by the cooperation between University of Kassel (UNIK) and the DLR Göttingen, in Germany [5].

Now the cooperation between the UNIK and CEA Saclay is set up to develop further the application of algorithm in civil engineering. By this cooperation, the research of TAMARIS can joint in the field of substructure testing and the UNIK can investigate more on the algorithm.

There are two main types of substructure tests which can be aimed to increasing capability of testing facilities. The first and common type is that tested parts are coupled at their top level with the simulation models via some actuators. With the second one, tested samples are excited by applying only vibration at the bottom. With existing shaking tables, TAMARIS should go in the second direction for develop and apply substructure testing. From this, RTSTs of TMDs are chosen as studied objects of the collaboration between CEA and UNIK.

2   DEVELOPMENT OF THE CONTROL SYSTEM IN CEA SACLAY FOR REAL-TIME SUBSTRUCTURE TESTING

The control system in TAMARIS can be developed for substructure testing on distributed tables. The shaking tables are controlled by a set of control devices and software. MTS controllers are working in closed loop with a typical time delay between the real response of the tables and the controlling signals. The frequency response of the system is also satisfied to the requirement of most of dynamic tests as well as RTSTs for civil applications.

In order to carry out RTSTs on one or two shaking tables, the algorithm of RTST has to be implemented in software to control the testing process. In principally, the requirement of the control software is not only to be able to control the movement of the shaking tables but also to measure fed back signals and to communicate with technician for input and output of test data. Some possibilities for setting up a control system to carry out RTST in TAMARIS have been discussed. In the case without any extra equipment, it is almost impossible to integrate directly a testing algorithm into the existing control devices for doing RTSTs on some shaking tables in a real time manner. A possibility is that an extra control device that can communicate to all existing controllers should be equipped for doing RTSTs. A real-time control unit such as ADwin system or MicroStar boards on PC and so one can be used for this purpose. In our project, ADwin system from the company "Jäger C.M. GmbH" (in Germany) was chosen for the application.

ADwin system is an advanced digital device for real-time testing and control. The ADwin hardware has a processor for control of its components to work as a scheduled process. With A/D converter, analog inputs such as coupling force or responses of the experimental substructure can be converted to digital data for processing in ADwin system. With D/A converter, digital control

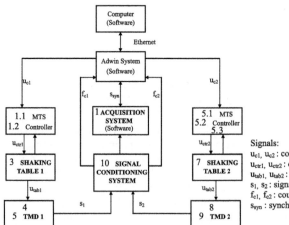

Figure 1.   Control diagram including ADwin and shaking tables.

signal such as displacement data of simulation substructure can be converted to analog signal for control of actuators or shaking tables. Beside these, the built-in memory in ADwin can be used for loading up to 10 process programs and for restoring temporary measured data. More over, ADwin can be connected to computer via an Ethernet or an USB connection. The control system for RTST on distributed shaking tables can be seen in figure 1.

It is well known that a high level program on computer with Windows operating system can not control scheduled commands exactly with an accuracy of millisecond. Unlike with multi tasking operating system as Windows, the software running in ADwin can control all commands and functions exactly as was planned in every clock event. Depending on the processor, the clock cycle in an Adwin can range from 3.3 ns to 25 ns. Of course, some tasks such as A/D or D/A conversion, data processing, etc. may take certain time durations about few or many clock cycles. From this, in order to be sure that the software can control time exactly, the duration for a control loop must be larger than the time which the CPU and other modules need to complete all tasks of the loop. In reality, most of RTSTs for real applications can be controlled by a high speed ADwin system such as ADwin Pro with T10 or T11 processor.

3 THE TESTED MODEL

The model consists of two parts as shown in figure 2. The main structure includes a frame of two storeys. The Tuned Mass Damper (TMD) part includes two TMDs and they are located at the second floor in figure 3. The mass of each TMD can vibrate in only one direction. The TMD 1 is assembled on the frame for vibration in a horizontal direction (OX) while the TMD 2 is used for the vibration in the other horizontal direction (OY).

At the first floor, the mass consists of two main steel masses with total mass of 1660 kg. These masses are bolted to the steel framework at the first floor. At the second floor, a steel frame can

1: AZALEE shaking table,
2: Frame of two stories,
3: Added mass at first floor
4: Rigid columns,
5: TMD 2,
6: TMD 1

Figure 2.   The frame with 2 TMDs.

1: Second floor of the frame, 2: Load cell for measuring coupling force of TMD 1, 3: TMD 1, 4: Load cell for checking other coupling forces, 5: load cell for measuring coupling force of TMD 2, 6: TMD 2

Figure 3.   Two TMDs on the second floor of frame.

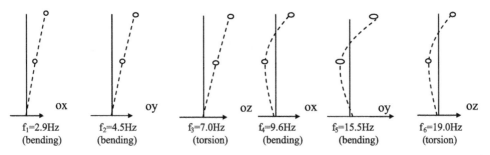

ox	oy	oz	ox	oy	oz
$f_1=2.9Hz$	$f_2=4.5Hz$	$f_3=7.0Hz$	$f_4=9.6Hz$	$f_5=15.5Hz$	$f_6=19.0Hz$
(bending)	(bending)	(torsion)	(bending)	(bending)	(torsion)

Figure 4.   Illustration of mode shapes of the main frame without extra mass or TMD.

carry two TMDs. The TMD 1 is placed on top while the TMD 2 is hanged to the bottom of the second floor. The concentrated mass of the second floor can range from 220 kg (without TMD) up to maximal value of 700 kg (with TMDs and extra mass) in order to change the frequencies of the model for differential investigations.

The main frame is designed for testing with excitations in two horizontal directions. The six first frequencies and the respective shape modes of the main frame can be seen in simple form in figure 4.

Table 1. Parameters of the two TMDs.

Substructure	Mass of the box m_b(kg)	Active mass m_s(kg)	Stiffness k_s(N/m)	Damping c_s(Ns/m)	Frequency range (Hz)
TMD1	80	37–95	≈18700	≈220	2.23 ÷ 3.55
TMD2	80	37–95	≈34300	≈270	3.03 ÷ 4.90

There are some notations on these frequencies. These frequencies are distinguishable clearly, thus, it is convenient for investigating on differential modes of vibration. However, there is a torsion mode (f_3) in the range of frequencies (from f_1 to f_4) of bending modes. To avoid complicated effects of torsion vibration when the model is being tested in vibration in two horizontal directions, the measurement should not sensitive to the torsion modes. It is important that these frequencies be located in a practical range of many applications in civil engineering as a planned purpose of the project.

To be accompanied with the frame, each TMD has a same range of frequency with the frame in respectively horizontal direction. The masses of TMDs could be changed in order to adjust the eigen frequencies according to the first bending modes of vibration of the frame (including mass of the boxes of TMDs and extra mass at the second floor) in two horizontal directions. The main parameters of two TMDs are listed in table 1.

4 SIMULATION OF MODEL AND RTST OF TMD AND PLAN OF TESTS

The full model in reference test under excitation of vibration of shaking table in a horizontal direction can be seen in figure 5. In RTST of a TMD with vibration in the horizontal direction, the numerical substructure would be treated as a two-DOF system in figure 6 and the TMD can be tested as figure 7.

The vibration equations of the frame without TMD and that of the simulation part subjected to a load due to vibration of shaking table can be described as equation (1) and (2) respectively.

The load of the model

$$\begin{bmatrix} m_{11} & m_{12} \\ m_{21} & m_{22} \end{bmatrix}\begin{bmatrix} \ddot{u}_1 \\ \ddot{u}_2 \end{bmatrix} + \begin{bmatrix} c_{11} & c_{12} \\ c_{21} & c_{22} \end{bmatrix}\begin{bmatrix} \dot{u}_1 \\ \dot{u}_2 \end{bmatrix} + \begin{bmatrix} k_{11} & k_{12} \\ k_{21} & k_{22} \end{bmatrix}\begin{bmatrix} u_1 \\ u_2 \end{bmatrix} = \begin{bmatrix} f_1 \\ 0 \end{bmatrix} \tag{1}$$

$$\begin{bmatrix} m_{11} & m_{12} \\ m_{21} & m_{22} + m_e \end{bmatrix}\begin{bmatrix} \ddot{u}_1 \\ \ddot{u}_2 \end{bmatrix} + \begin{bmatrix} c_{11} & c_{12} \\ c_{21} & c_{22} \end{bmatrix}\begin{bmatrix} \dot{u}_1 \\ \dot{u}_2 \end{bmatrix} + \begin{bmatrix} k_{11} & k_{12} \\ k_{21} & k_{22} \end{bmatrix}\begin{bmatrix} u_1 \\ u_2 \end{bmatrix} = \begin{bmatrix} f_1 \\ 0 \end{bmatrix} + \begin{bmatrix} 0 \\ f_c \end{bmatrix} \tag{2}$$

$$f_1 = (k_{11} + k_{12})u_0 + (c_{11} + c_{12})\dot{u}_0 \tag{3}$$

Where, m_{11}, m_{12}, m_{21}, m_{22} are values of the mass matrix of the steel frame; m_e is the effective mass of the other TMD on the second floor; k_{11}, k_{12}, k_{21}, k_{22} are values of the stiffness matrix of the frame; c_{11}, c_{12}, c_{21}, c_{22} are values of the damping matrix of the frame; f_1, f_2 are the loads on the full model; f_c is the coupling force between the experimental part and the numerical part.

The damping matrix C can be calculated from mass and stiffness matrices as Rayleigh damping as equation (4). The model in vibration in each direction is two-DOF system, thus, the coefficients α and β can be determined from two equations of two eigen modes as equation (5).

$$C = \alpha \cdot M + \beta \cdot K \tag{4}$$

$$\alpha + \beta \cdot \omega_i^2 = 2\omega_i \cdot \xi_i, \quad i = \overline{1,2} \tag{5}$$

For RTST of TMD, the model of frame should be simulated as exactly as possible. All of mass, damping and stiffness matrices of the frame should be identified. There are some techniques that cab

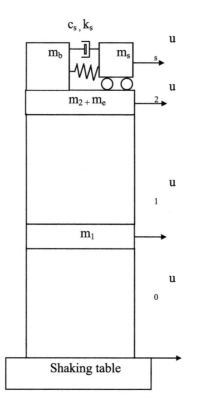

Figure 5.   Full model.

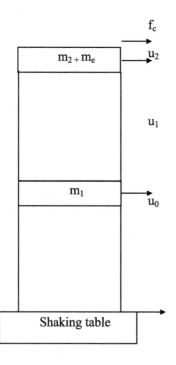

Figure 6.   Numerical model.

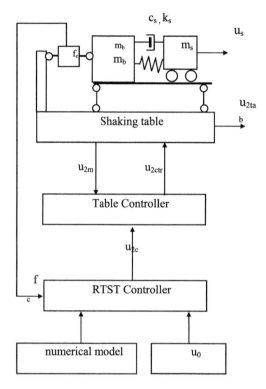

Figure 7.   RTST of a TMD.

be used to obtain relatively exact parameters for modeling of the real frame. A common technique is using the static condensation method to obtain the condensate mass and condensate stiffness matrices of the real frame. However, it is impossible to produce the exact full matrices of masses and stiffness for this real frame; it is needed to adjust the parameters further by investigating on dynamic characteristics of the model. As which was projected, the model is simple and its parameters can be identified from some dynamic tests. In this research, the parameters for modeling of the frame are measured directly and identified by using measured transfer functions.

In the RTSTs of TMDs, it is not needed to model TMDs. However, a reasonable model of TMD should be used for planning RTSTs. For this purpose, each TMD could be simulated in figure 7 as an ideal SDOF system with a dump mass (box of TMD), an active mass, a damper and a spring. It is assumed that the friction between the active mass and the box of TMD can be neglected. With this model and the assumption, the response of the full model including frame and the TMDs subjected to a vibration of shaking table in one direction can be computed theoretically. It should be mentioned that, the real response of structure in a tests should be smaller than that of simulation due to the effect of friction in TMDs.

Our example of the result of simulation of a RTST of TMD 1 with sine sweep load and a time interval of 5 ms, by using the control software could be seen in figures 8 and 9. Results from the simulation (pink and yellow graphs) and from the exact solution (blue graph) meet together.

In this research, for investigation and verification of the algorithm of RTST, a series of RTSTs of TMDs are going to be carried out with variation of combination of structure system (including simulation and experimentation parts), with differential loads (sine sweep and earthquake loads) in two horizontal directions in the table 3. Beside RTSTs of TMDs, series of reference tests in table 2 are also being conducted for validation of RTSTs. The tests of frame (No. 1.1.1 and 1.1.2) are conducted to identify real parameters of the steel frame. From dynamic response of the frame, the frequencies of TMDs are adjusted by carrying out tests No. 1.2.1 and 1.2.1 in order to work at the

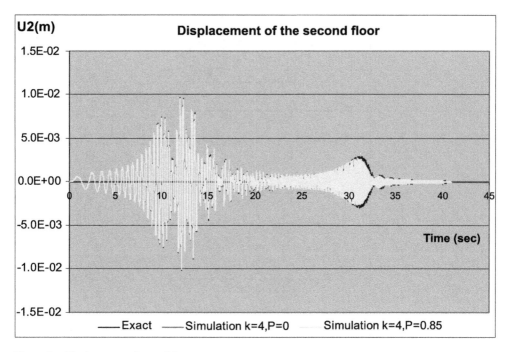

Figure 8.   Displacement of second floor.

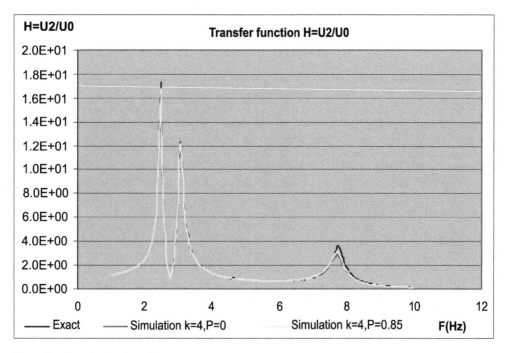

Figure 9.   Transfer function of system.

Table 2. Reference tests of the model.

Test no.	The tested structure	Direction	Loading type
1.1.1	Frame	X	Sine sweep
1.1.2	Frame	Y	Sine sweep
1.2.1	TMD 1	X	Sine sweep
1.2.2	TMD 2	Y	Sine sweep
1.3.1	Frame + TMD 2 (inactive)	X	Sine sweep
1.3.2	Frame + TMD 1 (inactive)	Y	Sine sweep
1.4.1	Frame + TMD 1 (active) + TMD 2 (inactive)	X	Sine sweep
1.4.2	Frame + TMD 2 (active) + TMD 1 (inactive)	Y	Sine sweep
1.4.3	Frame + TMD 2 (active) + TMD 1 (active)	X,Y	Sine sweep
1.4.4	Frame + TMD 1 (active) + TMD 2 (inactive)	X	Earthquake record
1.4.5	Frame + TMD 2 (active) + TMD 1 (inactive)	Y	Earthquake record
1.4.6	Frame + TMD 1 (active) + TMD 2 (active)	X,Y	Earthquake records

Table 3. Real time substructure tests of TMDs.

Test no.	Simulation part	Experimental part	Direction	Loading type
2.4.1	Frame + mass of TMD 2	TMD 1	X	Sine sweep
2.4.2	Frame + mass of TMD 1	TMD 2	Y	Sine sweep
2.4.3	Frame + mass of TMD 2	TMD 1	X	Sine sweep
	Frame + mass of TMD 1	TMD 2	Y	
2.4.4	Frame + mass of TMD 2	TMD 1	X	Earthquake record
2.4.5	Frame + mass of TMD 1	TMD 2	Y	Earthquake record
2.4.6	Frame + mass of TMD 2	TMD 1	X	Earthquake
	Frame + mass of TMD 1	TMD 2	Y	Records

Table 4. Planned parameters of RTSTs of TMDs.

Time step T_s (ms)	Sample rate f_s (Hz)	Number of substeps k_{sub}	Error compensation P
10, 20	100, 50	1, 2, 3, 4, 5	0, 0.85, 0.9, 1.0

first frequencies of frame in two horizontal directions. The tests No. 1.3.1 and 1.3.2 are conducted to check the parameters of the simulation part including frame and rigid mass for RTSTs. It is most important that, tests of full model including frame and TMDs (numbered from 1.4.1 to 1.4.6) be being carried out for verification of the RTSTs of TMDs (tests numbered from 2.4.1 to 2.4.6).

Concerning to the algorithm, there are many objectives that are being observed and verified from the RTSTs in this research. These are listed as specification of integration scheme in the algorithm; accuracy, stability and limitation of the algorithm with and without sub-stepping process; error development and error compensation mechanisms; feasibility of RTST at TAMARIS for applications in civil engineering.

In order to study on these objectives, it is need to carry out RTSTs of TMDs with series of parameters such as time interval of step and sub step, with or without error compensation. These variants are intended as in table 4.

5 DISCUSSION

The project actually started almost a year; however, a lot of works have been carried out. Firstly, a control system including equipment of TAMARIS and Adwin system have been developed and now available for performing RTST on distributed tables in CEA Saclay. Secondly, the model of steel frame and TMDs was produced for doing reference tests and RTSTs of TMDs. In addition, some first reference tests are being carried out on the model and it shows that the model is working as well as what was planned.

In the upcoming time, the planned RTSTs are going to be performed in CEA Saclay and series of objectives would be investigated. From this, the algorithm of RTST developed by UNIK is going to be discussed further for applications in civil engineering.

By conducting this project, CEA Saclay can join into field of RTST for potential applications in civil engineering. Moreover, with the collaboration between CEA Saclay and UNIK as well as other testing centers, the network of RTSTs in the world could be developed further.

REFERENCES

1. P. Sollogoub, J.C. Quéval, Th Chaudat (2003), Seismic Testing Capabilities of CEA Saclay Laboratory: Description—Evolution—Development International Collaboration, NCREE/JRC Workshop, Taipei, Taiwan.
2. Veit Bayer, Uwe E. Dorka, Ulrich Füllekrug, Jörg Gschwilm 2005, *real-time pseudo-dynamic substructure testing: algorithm, numerical and experimental results*. Aerospace Science and Technology 9 (2005), pages 223–232.
3. Uwe E. Dorka (2002), *Hybrid experimental—numerical simulation of vibrating structures*, Proceeding of The International Workshop WAVE 2002, pages 183–191. Okayama, Japan.
4. Uwe E. Dorka (1990), *Fast online earthquake simulation of friction damped systems*, SFB151 Report no. 10, Ruhr-University Bochum.
5. Uwe. E. Dorka, Ulrich Füllekrug (1998), Report of The DFG-project Do 360/7: Sub-PSD Tests, Germany.
6. Van Thuan Nguyen, Uwe E. Dorka (2006), Application of digital technique in a control system for real-time sub-structure testing, 4th World Conference on Structural Control and Monitoring, San Diego, USA (presented).

CHAPTER 13

Networked hybrid simulation of large-scale structures in Taiwan

K.C. Tsai & S.H. Hsieh
Department of Civil Engineering, National Taiwan University, Taipei, Taiwan
National Center for Research on Earthquake Engineering, Taipei, Taiwan

Y.S. Yang & K.J. Wang
National Center for Research on Earthquake Engineering, Taipei, Taiwan

ABSTRACT: An Internet-based framework, called the Internet-based Simulation for Earthquake Engineering (ISEE), was developed to facilitate networked hybrid simulations collaboratively performed by earthquake engineering laboratories in Taiwan or across the world. Two approaches were developed in this platform. One of the approaches, the Database Approach, employs a database server as a data exchange medium for inter-laboratory communications, offering an easy way to perform multi-site networked collaborative hybrid simulations. The other, the Application Protocol Approach, runs directly on top of Transmission Control Protocol/Internet Protocol (TCP/IP) and offers a network-time efficient way to run networked hybrid simulations. Both the approaches can work independently. Several networked hybrid simulations of large-scale structures using either approach are presented, showing the feasibility of the ISEE platform for domestic and international networked hybrid simulations.

1 INTRODUCTION

1.1 *Background*

The increasing complexity and scale of structural experiments are reflected in rising costs. As a result, existing earthquake engineering laboratories are gradually becoming incapable of satisfying the demands for various types of large-scale and complicated experiments. Rather than endlessly increasing the capacity of each laboratory, it would be more cost effective for geographically distributed earthquake engineering laboratories to collaboratively conduct networked hybrid simulations through a network system. In this concept, sub-components of a prototype structure, either a physical specimen or a numerical simulated component, can be tested at different laboratories, with the results disseminated through a communications network.

1.2 *Efforts on networked hybrid simulation*

Research efforts have been made to develop technology which allows multi-site collaborative experiments. In Japan, network techniques were applied to hybrid simulations such as a base-isolated building model (Pan et al., 2005). In Korea, a networked hybrid simulation of a base-isolated bridge was carried out collaboratively at three laboratories that are hundreds of miles apart from each other (Park et al., 2004). A network system for E-Defense, which is a full-scale earthquake testing facility in Japan, is being developed at present (Otani et al., 2003). In China, a system called NetSLab, was developed and has been used for bridge pier and pile foundation hybrid simulations (Guo et al., 2006). In Europe, the European Laboratory for Structural Assessment completed a single-site hybrid simulation by collaboratively using multiple networked computers (Pinto et al., 2004). In the United States a research project, called the Network for Earthquake Engineering Simulation (NEES), aims to explore the benefits of sharing and integrating laboratory resources including equipment, experiment data, and simulation software via networks. A variety of experiment facilities have been updated or constructed in the NEES project, and, a grid system called NEESgrid (Whitmore et al., 2006) has been constructed.

1.3   *Efforts in Taiwan*

The National Center for Research on Earthquake Engineering (NCREE) has developed a platform called Internet-based Simulation for Earthquake Engineering (ISEE) for collaborative hybrid simulation among laboratories through the Internet. Two approaches have been developed in the ISEE platform: the Database Approach and the Application Protocol Approach. Each approach can work independently and can be employed to conduct a networked collaborative hybrid simulation. The Database Approach uses a database server for inter-laboratory communications. It is easy to monitor the experimental data and progress as well as to tailor additional programs to expand the functions for a networked experiment by accessing the database. In the Application Protocol Approach, a protocol called Networked Structural Experiment Protocol (NSEP) has been proposed, which would run directly on top of the Transmission Control Protocol/Internet Protocol (TCP/IP). Based on relatively low level network protocol and well-defined communication protocol, this approach offers a way of running networked hybrid simulations that use network time very efficiently. The ISEE platform has been applied to networked single-site and multi-site hybrid simulations, using either the Database Approach or the Application Protocol Approach.

2   FRAMEWORK OF THE DATABASE APPROACH

The aim of the Database Approach is to offer an easy-to-use software platform for networked hybrid simulations. A networked hybrid simulation requires network interconnection among geographically separated participating programs to allow for the exchange of analytical and measured data continuously during the experiment. These programs could include an Analysis Engine, which simulates the numerical parts of the tested structure, and one or more Facility Controllers, which control the hydraulic actuators that deform specimens. Each hybrid simulation normally has a unique setup of the Analysis Engine, Facility Controllers, specimens, and the interconnections among them. The Database Approach simplifies the tedious work of modifying, rewriting and testing the source codes of these participating programs for a networked hybrid simulation. A brief introduction of the Data Center, Analysis Engine and the Facility Controller follows. More details can be found in Yang et al. (2007).

2.1   *Data Center*

The Database Approach eases the complexity by employing a database server to act as a Data Center. This allows authorized geographically distributed programs to input and extract analytical data, measured data, and other user-defined information which can be expressed by floating-point numbers. All participating programs communicate only with the Data Center (see Figure 1) using the widely used SQL database access protocol—direct communication among the participating programs is not needed. In addition to serving as a database, the Data Center also manages the interconnection relationships among participating programs for each hybrid simulation.

The Data Center stores three data tables for each hybrid simulation: an experimental data table, a data description table and an interconnection table. An example of a four-pier bridge experiment (see Figure 2) showing how the three tables are used is given below. Two middle piers are physical specimens located at two separate laboratories, and the remaining components are numerically simulated by an Analysis Engine. The data description table (Table 1) contains descriptions of the eight columns in the experimental data table (Table 3). Table 2 determines the data input and retrieval columns for each participating program. The experimental data table stores the experimental data and also shows the progress of the experiment. Table 3 contains the data of

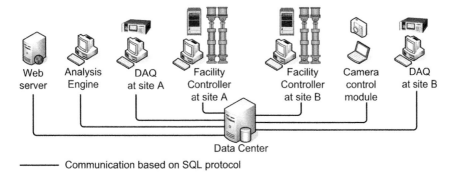

Figure 1.   Database Approach in ISEE.

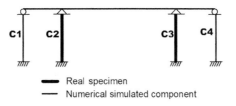

Figure 2.   A hybrid simulation example: A 4-column bridge with two pier specimens.

Table 1.   A data description table.

Table column	Description	Unit
1	Displacement along X of pier C2	mm
2	Displacement along Y of pier C2	mm
3	Displacement along X of pier C3	mm
4	Displacement along Y of pier C3	mm
5	Resisting force along X of pier C2	kN
6	Resisting force along Y of pier C2	kN
7	Resisting force along X of pier C3	kN
8	Resisting force along Y of pier C3	kN

displacements. This table indicates whether the Facility Controllers have uploaded the measured forces into the Data Center or if they are still either deforming the specimens or accessing the Data Center.

2.2   *Analysis Engine*

An open-source structural analysis program, called OpenSees (Fenves et al., 2001), has been adopted and modified for use in the Database Approach. A new element class named *pseudoGens*, derived from the Element class in the OpenSees framework (Figure 3), has been implemented. A *pseudoGen* element can contain an arbitrary number of nodes. Users need only correctly add *pseudoGen* elements into the OpenSees input file. In each time step in the Analysis Engine, the predicted displacements are calculated and then each *pseudoGen*, which represents a specimen, receives a request to return resisting forces corresponding to the predicted displacements. *The pseudoGen then sends the displacements to the Data Center, waits and gets the corresponding (static) resisting forces, combines them with inertia and damping forces and returns the (dynamic) resisting forces to the OpenSees dynamic integration.*

Table 2.   An interconnection table.

Participating program	Program ID	Table columns to put	Table columns to get	Programs to wait
Analysis Engine	1	1 2 3 4	5 6 7 8	2 3
Facility Controller of pier C2	2	5 6	1 2	1
Facility Controller of pier C3	3	7 8	3 4	1

Table 3.   An experimental data table.

Time step	Table columns							
	1 (mm)	2 (mm)	3 (mm)	4 (mm)	5 (kN)	6 (kN)	7 (kN)	8 (kN)
1	3.13e-3	−3.24e-3	3.76e-3	4.56e-3	0.1221	−1.8017	−2.4037	−2.4420
2	1.13e-2	−0.52e-3	1.09e-2	1.41e-2	−0.0305	−0.4883	−2.4045	−2.2904
3	2.36e-2	4.54e-3	2.42e-2	2.74e-2	(empty)*	(empty)	(empty)	(empty)
....**								

*The empty cells show that the Data Center is waiting for the resisting forces data from the Facility Controllers.
**A typical hybrid simulation runs thousands of time steps.

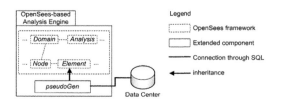

Figure 3.   Framework of the OpenSees-based Analysis Engine.

Table 4.   Usage of *pseudoGen* element.

```
element pseudoGen $DIP   $TID   $PID   $ETag   $Nn   $N1   $N2 ......  $Filename
    # $DIP: IP address of the Data Center.   # $TID: Test ID.   # $PID: Program ID.
    # $ETag: Element tag number.   # $NN: Number of adjacent nodes.   # $N1 $N2 ...: Node IDs
    # $Filename: File name of the mass and damping matrices
```

A PISA3D-based Analysis Engine has also been developed using a similar approach to the OpenSees-based one. Further details of the PISA3D-based Analysis Engine can be found at Tsai and Lin (2003). An Analysis Engine can be replaced by a general command-generation module. For example, a predefined displacement command generator can be used for pushover or cyclic tests. It can be used to run series pushover or cyclic tests after a hybrid simulation experiment.

2.3   Facility Controllers

A Facility Controller is a software component used to control an experimental facility in a laboratory. In the Database Approach, a Facility Controller gets displacements from the Data Center, deforms the specimens according to the displacements and sends the measured resisting force back to the Data Center. Facility Controllers are hardware dependent. At present, Facility Controllers for MTS FlexTest IIm, MTS 407 and MTS 458 have been developed.

3 A SIMULATION EXAMPLE USING ISEE DATABASE APPROACH

This section demonstrates a transnational networked hybrid simulation between Taiwan and Canada using the Database Approach of ISEE. This simulation is called DSCFT-2006.

3.1 *Brief description of DSCFT-2006 hybrid simulation*

The DSCFT-2006 simulated the dynamic responses of a five-span double-skinned-concrete-filled tubular (DSCFT) bridge subjected to a series of ground motions. Three of the four DSCFT piers were physical scaled down specimens and were separately located at NCREE, National Taiwan University (NTU) in Taiwan and Carleton University (CU) in Canada. The remaining parts were numerically simulated by either an OpenSees-based or a PISA3D-based Analysis Engine in different tests. Figure 4 shows the network configuration of the experiment.

The participating programs in the experiment include an Analysis Engine, NCREE Facility Controller, NTU Facility Controller, CU Facility Controller, NCREE Data Acquisition, NTU Data Acquisition, and a data transformer. In addition, a GISA3D-based visualization module (Tsai and Chuang, 2005) generates the dynamic animations of the bridge using 3D computer graphics. A web server instantly broadcasts real-time videos of the specimens and that of the bridge responses through 2D plots and a 3D view generated by the visualization module. A camera module keeps track of the progress of the experiment through the Data Center and automatically takes pictures of the NCREE specimens every other time step. Further information about the DSCFT-2006 can be found at Tsai et al. (2007). Experimental verifications are available at http://exp.ncree.org/dscft.

3.2 *Timing statistics*

In the main three-site hybrid simulation tests in the DSCFT-2006 experiment, each time step costs about three to four seconds. Since each time step represents 0.02 second in analysis, the experimental time cost on average is about 150 to 200 times of the analytical time. However, the time cost is not quite constant. The longest elapsed time for a single time step may be up to 17 seconds mainly due to the time cost of facility control. Table 5 shows two of the ground motions tests in the DSCFT-2006 simulation.

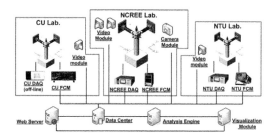

Figure 4.   Network connectivity of the DSCFT-2006 simulation.

Table 5.   Time cost of each time step in the DSCFT-2006 simulation.

	Analysis Engine	NCREE Facility Controller	NCREE DAQ	NTU Facility Controller	NTU DAQ	CU Facility Controller	Network & Data Center	Total
Average	0.06 s	1.32 s	0.89 s	1.03 s	0.81 s	0.94 s	0.68 s	3.02 s
Max.	0.36 s	6.66 s	0.90 s	1.80 s	1.31 s	16.01 s	1.66 s	17.45 s

4   FRAMEWORK OF THE APPLICATION PROTOCOL APPROACH

4.1   *Platform for Networked Structural Experiments (PNSE)*

A computer networking platform, called the Platform for Networked Structural Experiments (PNSE), and its associated application protocol (AP), called the Networked Structural Experiment Protocol (NSEP), were developed to demonstrate the Application Protocol Approach. In the Application Protocol Approach, the communication is implemented by directly using TCP/IP. PNSE is a reliable platform for robust progression of networked testing. It is a multi-client computer networking system with each client connected to the server via a point-to-point TCP connection channel. Figure 5 illustrates the architecture of PNSE. Communication between a client and the server is achieved by sending and receiving data packets that are well defined by an associated AP. Two kinds of clients, the Command Generation Module (CGM) and the Facility Control Module (FCM), were defined on the PNSE. The CGM generates the commands to be imposed on specimens by all participating FCMs. In each command execution loop (called a 'command cycle' in this study) the commands are generated by the CGM and sent to the server, which then dispatches the commands to all participating FCMs. After executing the command and measuring and/or calculating the specimen responses, the FCMs will send the response data back to the server, which then forwards this data to the CGM to complete one command cycle. The command cycle is then repeated until the test ends.

Signals on the PNSE refer to values that change continuously with respect to time. Typically they include the structural responses obtained in labs and the commands generated by the CGM. Signals are classified as either "critical signals" or "open signals". Critical signals are those that are prerequisite for the progression of testing. For example, the computed target displacement and the measured restoring forces in a hybrid simulation test scenario are critical signals because, if any of them fail to reach their destination, the whole experiment would be suspended. Open signals, on the other hand, are data values that have been designated for near real-time broadcast on the Internet; for example, strain gauge values. Even without transmitting them in each command

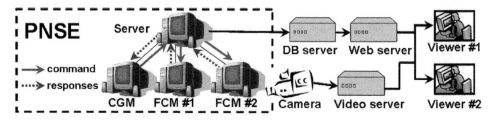

Figure 5.   The architecture of PNSE.

Table 6.   Data packets that are currently defined in NSEP.

Type	Usage
ERROR	Error notification
LOGIN	Client name/password for login procedure
EXPINFO	Experimental meta-data
EXPSTATE	Current test running state
CLNSTATE	Current running state of a specific client
CSIG_CGM	Critical values (between the CGM and the server)
CSIG_FCM	Critical values (between the FCM and the server)
OSIG_SC	Open signal values from the server to all clients
OSIG_CS	Open signal values from a client to the server
DISCUSS	Human communication (human readable text)

cycle, the networked test can still advance to the next step. Nevertheless, typically open signals are carefully chosen and broadcast in each command cycle since they facilitate monitoring of the test and understanding of the specimen behavior.

In networked testing, human to human communication is still necessary because it is difficult to comprehensively define all possible events that may occur in a structural experiment, within the AP. Typical examples include the detailed damage condition of the specimen and the corresponding actions taken to fix it. The PNSE addresses this issue by defining a data packet type in the associated AP that contains readable text to allow direct human communication.

In the course of an experiment, the test may be suspended or stopped prematurely for a variety of reasons. Due to laboratory safety, all FCMs have the authority to autonomously change their individual running state, although they are required to report the change to the system.

4.2   Networked Structural Experiment Protocol

The NSEP defines the following in detail: (1) the syntax of a data packet; (2) general information about the structural experiments; (3) significant laboratory events; and (4) signals, including commands and responses. The NSEP also stipulates the behavior of the server and the client on receiving a specific data packet. Detailed definitions of data packets can be found in the reference Tsai et al. (2003). Currently defined NSEP data packets are listed in Table 6. As for the mechanism of information transmission, NSEP stipulates that all PNSE programs make active notifications (AN). AN means that the one who owns the relevant information should actively inform the other side of the connection when appropriate. For example, it is the CGM's responsibility to actively send out the command, instead of being queried by the server.

4.3   Characteristics of PNSE

4.3.1   Environment independent
The PNSE is an environment independent platform since it is constructed based on the widely supported TCP/IP protocol suite. High interoperability can be assured regardless of heterogeneous environments due to diverse types of computer hardware, operating systems, programming languages, and facility controllers. This suggests that minimum programming effort will be required to incorporate existing programs running in these diverse environments.

4.3.2   Event-reflective
As mentioned earlier, in addition to the prerequisite commands and responses, significant laboratory events such as changes to the running state of FCMs are also adequately defined in NSEP. Combining this with the stipulation of AN, PNSE becomes an event-reflective platform, which means a platform with a formal mechanism that allows events that occur in a participating laboratory during a collaborative test to be transmitted to all other participating entities. Thus, all participants can be kept aware of the overall running condition of the test.

4.3.3   Efficient data transmission
An approach that is contrary to AN but commonly used to transfer messages is the polling mechanism (PM). It works by having the entity that requires relevant information actively and repeatedly query for the information until the information is acquired. PM has been widely adopted in other research work due to its simplicity. However, it generally decreases the overall system efficiency because the polling has to be performed sufficiently often such that the information is acquired promptly. It wastes a lot of time and network resources as the information queried is usually not available at the time of the query. On the other hand, using AN greatly increases the system's efficiency since all information is transmitted automatically to the destination without any time and network resources wasted for continuous polling.

4.3.4   *Simplified communication and high extensibility*

All PNSE clients connect only to the server, rather than to the other clients. As such, all data packets are transmitted through the server to their final destinations. Having this detour consumes more time in data transmission but the advantage is that it tremendously simplifies the network communication flow since each client program only has to deal with the server. This also translates into high extensibility of the platform as the design saves a large amount of programming effort, especially when events and interactions are complex.

4.3.5   *High flexibility to support versatile test configurations*

NSEP does not concretely define the content for critical signals since the content is different from case to case. For example, mapping between structural DOF and actuator DOF always changes with each test configuration. NSEP defers the determination of the contents of CSIG_CGM and CSIG_FCM to the participating researchers during the planning stage. This equips users with high flexibility in using PNSE to achieve their specific testing goals. In addition, signals are classified to allow different operations. For example, this is especially useful when the integration time step has to be short enough due to stability or accuracy issues, while equally dense measurements of specimen behavior are not necessary or are not affordable.

4.3.6   *High resuming ability after accidental disruptions*

In the course of an experiment, there could be instances of accidental disruptions that either temporarily or permanently interrupt the test. Examples include network connection failure, program run-time errors, or damage that occurs to unexpected parts of the specimen. If all the specimens remain elastic when an accidental disruption occurs, all that needs to be done is re-conduct the test again from the beginning after the problem is solved. However, if some specimens have entered their inelastic zones, re-conducting the test from the beginning again would not be ideal as this would cause all the specimens, including those that have no problems, to undergo erroneous plastic deformation. PNSE elaborates on experimental running states to avoid forced disconnection of all those clients that run smoothly and therefore prevents erroneous plastic deformation that might be experienced in those clients (Wang et al., 2007).

5   EXPERIMENTAL VALIDATION OF APPLICATION PROTOCOL APPROACH

To investigate the feasibility and data transmission efficiency of the PNSE, a series of transnational hybrid simulation tests were conducted on a two-pier bridge system. Figures 6 and 7 show the numerical model and photographs of the specimens. Details of the specimen design and testing parameters can be found in the reference Tsai and Yeh (2003). The two piers were fabricated and tested in the NCREE laboratory (NCREELab) and the National Taiwan University laboratory (NTULab), with FCMs programmed using C++ and LabVIEW, respectively. Three hybrid simulation tests (Tests A, B, and C), were conducted with all testing parameters exactly the same except for the locations of the PNSE server and the CGM program. In Test A, both the server and the CGM ran on the same computer in NCREE. In Test B, the server program ran at NCREE and the CGM ran at

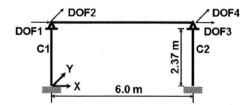

Figure 6.   Model of DSCFT hybrid simulation.

Stanford University. In Test C, the server resided at Stanford University and the CGM at NCREE. It was obvious that Test C, where all packets had to detour to the United States before reaching their destination in Taiwan, would be the most challenging one. Figure 8 shows the displacement commands sent and received by the CGM and the FCM, as well as the comparison between the experimental and numerical results. It can be seen that the signals were correctly transmitted, and that the test results matched those obtained with the simulation.

Figure 7.   Photographs of the DSCFT piers.

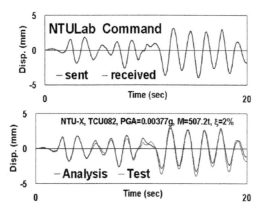

Figure 8.   Results of DSCFT hybrid simulation test.

Table 7.   Working time and round trip time measured on PNSE.

Activity		Test A	Test B	Test C
WT^*_{CGM}		0.0180	0.0060	0.0156
RTT^{**}_{CGM}	PNSE	0.0080	0.1661	0.1693
	Ping	<0.001	0.1628	0.1653
WT_{Server}		0.0437	0.0096	0.0459
WT_{NTULab}		0.5512	0.5486	0.6071
RTT_{NTULab}	PNSE	0.0037	0.0056	0.1601
	Ping	<0.001	<0.001	0.1625
$WT_{NCREELab}$		0.6356	0.6106	0.6540
$RTT_{NCREELab}$	PNSE	0.0046	0.0039	0.1701
	Ping	<0.001	0.0070	0.1608
T_{Total}		0.6634	0.7963	1.0545

*WT_x: working time of module x.
**RTT_x: round trip time between module x and the server.

The data transmission efficiency was measured by investigating the timestamps recorded in each command cycle. Table 7 lists the average time consumed on PNSE, and the time needed for the Microsoft Ping program, which was executed under similar network conditions for comparison purposes. For Test C, the round-trip time (RTT), that is, the time for a packet to travel from client to server and then back to the client, was 0.1701 second. Comparing this to the result of 0.1608 second obtained by the Ping program suggests that data transmission on PNSE was efficient as it was almost as fast as that provided by the operating system. Compared to conventional hybrid simulation tests, the additional time required to transmit data across the Internet for each integration step is only 0.3394 second, which is quite acceptable. Results of Test B were also compared to Experiment 153 in the Database Approach, where both the CGM and the Analysis Engine ran at Stanford University (Yang et al., 2007). The PNSE exhibited much better working efficiency as the average time for data transmission and data processing by the Application Protocol and Database Approaches were about 0.1813 and 1.944 seconds respectively (Wang et al., 2007).

6   COMPARISONS WITH THE DATABASE APPROACH

Comparisons between the current platforms developed by the two approaches are given in Table 8. Before further research results become available, the authors suggest using the Database approach if the higher cost of initial development for the Application Protocol Approach is not affordable and if the somewhat limited functionalities and poorer working efficiency of the Database Approach is acceptable. Otherwise, the Application Protocol Approach is recommended since it provides more comprehensive features and better working efficiency, which is needed for modern collaborative testing.

Table 8.   Comparisons between the current platforms developed by the Database and Application Protocol approaches.

Comparative item	Database Approach	Application Protocol Approach
Initial development	Easier	More difficult
Subsequent maintenance	Easier	More difficult
Supported OS	Microsoft	Microsoft, FreeBSD
Supported FEM	PISA3D, OpenSees	PISA3D
Transmitted content	Freely defined by the users	Clearly defined in NSEP
Supported transmitted data types	Floating-point numbers	Various types defined in NSEP
Information transmitted during tests	Signals*	Signals*, running states, human communication
Method to acquire information	Polling mechanism	Active notification
Event-reflective	No	Yes
Truly cooperative	Yes	Yes
Event-driven	No	Yes
Data transmission efficiency	Acceptable	High
Server-client communication	Simpler	More complex
Resuming ability	Low	High
Flexibility to support versatile test configurations	High	High
Extensibility to support new laboratory events	Low	High
Programs that need creation or modification to support new clients	The new client	The new client and the server

*Signals in Database Approach are all "critical", while in Application Protocol Approach they are classified as "critical" or "open".

7 CONCLUSIONS

Two different approaches, called the Database Approach and Application Protocol Approach, were developed to achieve the goal of networked collaborative structural testing. The Database Approach leverages off database technology to conduct networked collaborative experiments by using the SQL protocol. In addition, selected experimental data can be stored in the database, and consistency and durability of the experimental data can be assured. The Database Approach was employed in experiments conducted upon a DSCFT bridge model and an RCS frame and the feasibility and scalability of the Database Approach was well demonstrated and verified. The Application Protocol Approach established a multi-client computer networking system, PNSE, and its supporting AP, NSEP, to achieve the goal of networked collaborative testing. In addition to reliable progress of the structural experiments, PNSE also supports transmission of relevant information such as general experimental information, signals, significant laboratory events and human to human conversation. It exhibits satisfactory working efficiency and the characteristics of environment independence, being event-reflective, maximized extensibility and flexibility, and superior resuming capability after accidental interruptions, as verified by the series of transnational hybrid simulation tests conducted upon the DSCFT model. This paper closes with a detailed comparison between the two approaches. It can be concluded that both the approaches serve appropriately in networked collaborative structural testing.

ACKNOWLEDGEMENT

Financial support by National Science Council (NSC91-2711-3-319-200) and laboratory support provided by NCREE and NTU are acknowledged. Special thanks are also due to Mr. Bo-Zhou Lin and Mr. Ming-Chieh Chuang for their great efforts on PISA3D and GISA3D support.

REFERENCES

Fenves, G.L., Filippou, F.C., & McKenna, F. The OpenSees software framework for earthquake engineering simulation. *Proc. 2001 Structures Congress and Exposition*, Denver, CO, USA, 4–6 April 2002. (Available at http://www.curee.org/conferences/CP/docs/CP-proceedings.pdf)

Guo, Y.R., Xiao, Y., Fan, Y.L., Dong, X.H., & Hu, Q. 2006. Development and application of a collaborative hybrid dynamic testing software based on NetSLab. *Proc. 8th US natl. conf. on earthquake engineering*: paper no. 1198, San Francisco, CA, USA, 18–22 April 2006.

Ohtani, K., Ogawa, N., Katayama, T., & Shibata, H. 2003. Project E-Defense 3-D full-scale earthquake testing facility. *Proc. Joint NCREE/JRC workshop intern. collaboration on earthquake disaster mitigation research*: 223–234, Taipei, Taiwan, 17–18 November 2003. (Available at http://www.ncree.org.tw/ncree-jrc/CD/Schedule.htm)

Pan, P., Tada, M., & Nakashima, M. 2005. Online hybrid test by Internet linkage of distributed test-analysis domains. *Earthquake Engineering and Structural Dynamics* 34: 1407–1425.

Park, D.U., Yun, C.B., Lee, J.W., Nagata, K., Watanabe, E., & Sugiura, K. 2004. On-line pseudo-dynamic network testing on base-isolated bridges using Internet and wireless Internet. *Experimental Mechanics* 45: 331–343.

Pinto, A.V., Pegon, P., Magonette, G., & Tsionis, G. 2004. Pseudo-dynamic testing of bridges using non-linear substructuring. *Earthquake Engineering and Structural Dynamics* 33: 1125–1146.

Tsai, K.C. & Lin, B.Z. 2003. User manual for the platform and visualization of inelastic structural analysis of 3D systems PISA3D and VISA3D. Technical report CEER/R92-04, Center for Earthquake Engineering Research, National Taiwan University, Taipei, Taiwan.

Tsai, K.C. & Chuang, M.C. 2005. Development of an object-oriented graphical user interface for the structural analysis program. Technical report NCREE-05-012, National Center for Research on Earthquake Engineering, Taiwan.

Tsai, K.C., Yang, Y.S., Wang, K.J., Wang, S.J., Lin, M.L., Weng, Y.T., Chen, P.C., Cheng, W.C., Chuang, M.C., Lin, B.Z., Lau, D.T., & Chang, Y.Y. 2007. Network platform for structural experiment and analysis (ii): framework, protocol, software development for a Taiwan-Canada collaborative pseudo-dynamic experiment, Technical report NCREE-07-018, National Center for Research on Earthquake Engineering, Taipei, Taiwan.

Tsai, K.C., Hsieh, S.H., Yang, Y.S., Wang, K.J., Wang, S.J., Yeh, C.C., Cheng, W.C., Hsu, C.W., Huang, S.K. 2003. Network platform for structural experiment and analysis (i). Technical report NCREE-03-21, National Center for Research on Earthquake Engineering, Taipei, Taiwan.

Tsai, K.C. & Yeh, C.C. 2003. Networked substructure pseudo-dynamic tests of double-skinned CFT bridge piers under bi-directional earthquakes (1). Technical report NCREE-03-021, National Center for Research on Earthquake Engineering, Taipei, Taiwan. (In Chinese)

Wang, K.J., Tsai, K.C., Wang, S.J., Cheng, W.C., & Yang, Y.S. 2007. ISEE: internet-based simulation for earthquake engineering part II: application protocol approach. *Earthquake Engineering & Structural Dynamics*. (Accepted. DOI: 10.1002/eqe.729)

Whitmore, S., Van Den Einde, L., Warnock, T., Diehl, D., Hubbard, P., & Deng, W. 2006. NEESit software overview: IT tools that facilitate earthquake engineering research and education. *Proc. 8th US natl. conf. on earthquake engineering*: paper no. 966, San Francisco, CA, USA,18–22 April 2006.

Yang, Y.S., Hsieh, S.H., Tsai, K.C., Wang, S.J., Wang, K.J., Cheng, W.C., & Hsu, C.W. 2007. ISEE: internet-based simulation for earthquake engineering part I: database approach. *Earthquake Engineering & Structural Dynamics*. (Accepted. DOI: 10.1002/eqe.730)

CHAPTER 14

Real time substructuring in the UK

P.A. Bonnet
Science and Technology Facilities Council, Rutherford Appleton Laboratory, UK

A. Blakeborough, M.S. Williams
Department of Engineering Science, University of Oxford, UK

C.A. Taylor
Department of Civil Engineering, University of Bristol, UK

ABSTRACT: Real time substructuring simulates a building by splitting it into two components, a physical part, which ideally should be modeled at or near full scale, and a numerical part which simulates the remaining structure. The two components are connected together in a control loop. When the method simulates the behaviour of the building without a distortion in the timescale of the test it is termed real time substructuring. The success of the method depends on the efficient and accurate execution of the individual control loop tasks. A 3 DOF mass-spring rig is described which was used to compare the range of control algorithms used to improve the performance of the system. Then a test using a less usual form of the control loop is described with some outline results. Both sets of tests show that real time simulations can be successfully achieved with natural frequencies of the system up to 10 Hz.

1 INTRODUCTION

Laboratory testing of large structures has given rise to various test methods. For the analysis of structural response to complex dynamic loading, such as earthquake records, two basic experimental methods have co-existed for some time: the shaking table method, which is a fully dynamic experimental approach, and the pseudo-dynamic scheme, which applies the inertial forces to a full model of the structure under consideration. Hybrid simulation is an extension of the pseudo-dynamic scheme in which the structure to be modeled is split into two substructures. On one hand, there are the parts and regions that can easily be modeled numerically, either because they have a simple behaviour or because they are not considered to be critical for the analysis conducted. This is the numerical substructure. In contrast to this, there are the parts and regions of most interest, which must be realized physically, either because they are critical to the safety and performance of the structure or where perhaps a high degree of non-linearity is expected. This is the physical substructure.

The parts are connected in a control loop. The conventional method is to feed the earthquake excitation into the numerical model, and calculate the displacement of the nodes that are connected to the physical model. Actuators apply this displacement to the physical part and the induced forces are measured and then applied to the numerical substructure with the next component of the earthquake. If the physical substructure has significant velocity or acceleration dependent properties such as inertia, damping or strain-rate effects, the test should be carried out in real-time, that is to say with a time scaling factor of unity. This is real time substructure simulation, and follows our preferred definition of 'real-time', in contrast to the computer scientist's definition, which usually means interactive. An important benefit of this testing method lies in the possibility of its real-time character. There is, however, an implication of the motivation behind a real-time simulation that affects almost everything about the process. The purpose of a real-time test is

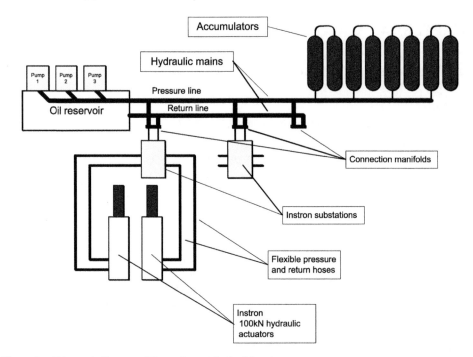

Figure 1.   Schematic diagram of the equipment in the laboratory.

to satisfy the need of the experimenter to capture some time dependent effect in the physical substructure that cannot be reproduced if the test were to be performed slowly. Achieving this imposes severe demands on the all components of the control loop since each part of the system must perform accurately under extreme time pressure.

We outlined our initial forays into this area in a review paper (Blakeborough et al., 2001). The rest of the chapter will outline some of the recent work at Oxford investigating various techniques to improve and control the physical and computational components as they perform towards their limits. The first piece of work investigates ways of improving the performance of the actuators, in particular methods to compensate for the phase lag or time delay introduced by the control system and the hydraulic jacks. This part was performed in collaboration with colleagues at Bristol. Further work regarding the performance of different algorithms to solve the numerical model is also referenced. Finally, there is a more detailed—but still brief—description of a set of real-time tests on a hysteretic damper that were conducted as part of the NEFOREEE (New Fields of Research in Earthquake Engineering Experimentation) European Community research program (Molina et al., 2006).

2   LABORATORY FACILITIES

The tests were performed in the Structural Dynamics Laboratory in the Department of Engineering Science at Oxford University, and to put the work into context a brief description of the laboratory and the testing equipment is necessary. Figure 1 is a diagram showing the layout of the equipment in the laboratory. Three pumps supply pressurized oil to a trunk main which is also connected to a bank of hydraulic accumulators. Actuators are connected to the oil supply by hydraulic hoses from the Instron sub-stations. The tests described in this chapter use 100 kN and 10 kN Instron® actuators. The 100 kN actuators are fitted with twin spool valves to give a high dynamic response.

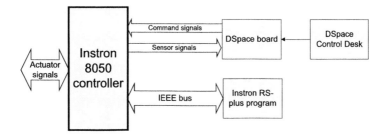

Figure 2.   Schematic diagram of the control and monitoring equipment.

The arrangement and connections between the control equipment are shown in figure 2. The actuators are controlled by an Instron® Series 8500 controller, which is itself regulated through an IEEE interface with proprietary software, RS-Plus, running on a PC. The outer control loop is run on a dSpace® digital signal processing board, housed in another PC. Simple control models can be programmed with Simulink® and more complicated systems and functions can be programmed in C. The board has access external ADC and DAC ports connected to terminals on an extension board. These are connected to the monitoring signals and the force and displacement demand and feedback signals from the Instron® controller via BNC connectors.

3   THREE DOF MASS-SPRING RIG TESTS

A successful real-time test depends on the quality of the performance of the actuators and their associated control system and the efficient integration of the numerical substructure equations of motion. Even a well-tuned actuator system will not deliver perfect performance. There will always be errors in the timing and amplitude of the imposed displacements (Bonnet et al., 2005). The timing error can render a real-time hybrid experiment unstable by introducing a phase lag or time delay into the system which constantly injects more energy into the system than is dissipated (Horiuchi et al., 1996). There are basically two approaches to compensating for this lag/delay. The traditional control solution is to put a lag compensator into the system. This is the basis of the MCS (Stoten, 1993) approaches, which are in effect tunable compensators. The other more pragmatic approach is to assess the time delay between the command and the response in the actuator and apply a forward prediction of the command signal to compensate (Darby et al., 2002, Wallace et al., 2005). The advantage of the second class is that it is easier to tune them adaptively during the test. At heart both strategies perform the same service; they are both forward filters attempting to compensate for delays in the execution of the command signal.

We have performed a large series of tests to compare the various compensation algorithms (Bonnet et al., 2006). Those considered are the MCS family, the Darby estimator, the modified Darby estimator (Bonnet, 2006), exact polynomial fitting extrapolation Horiuchi et al. (1996), least squares polynomial fitting extrapolation (Wallace et al., 2005), extrapolation based on linearly predicted acceleration, Horiuchi & Konno (2001), the Laguerre extrapolator (Bonnet, 2006) on linear systems and for non-linear gap spring systems, (Bonnet et al., 2007). There has also been a similar series of tests comparing the different numerical integration algorithms, all implemented for consistency on the dSpace® board (Bonnet, 2006, Bonnet et al., 2007b). Typical of these tests is the experiment described below.

Figure 4 shows the results from the emulation test in which the system was excited by a sine sweep covering the frequency range 0 to 10 Hz. The upper plot gives the overall power spectrum of the displacement of mass #2, which shows the two natural frequencies of 3.3 Hz and 7.9 Hz. The lower trace plots the envelope of the response trace as a function of time. The errors are greatest near resonance where the displacements are greatest, but the errors are always relatively small.

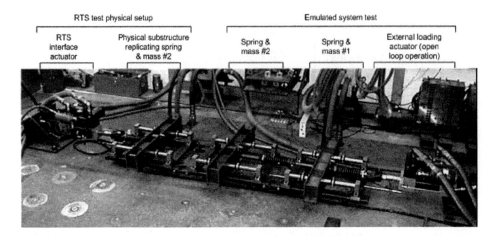

Figure 3.   Three DOF mass/spring test rig.

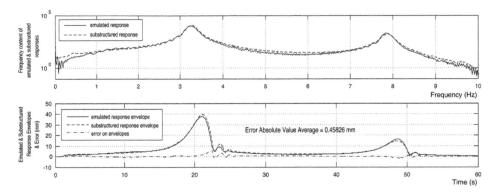

Figure 4.   System response plots for displacement of mass #2, upper—spectral amplitude, lower—envelope during sine sweep.

The reason for the disparity is that the natural frequencies of the two systems are not exactly the same and the magnification factor for each rig will be different, especially near resonance.

4   NEFOREEE TEST PROGRAM

This section describes testing performed as part of the European Communities NEFOREEE program. These real-time hybrid tests were linked with companion pseudo-dynamic tests at JRC Ispra and shaking table tests at NTUA. An outline of the main findings of the collaboration has been published (Molina et al., 2006). A brief description of the testing methods adopted for the Oxford work will be given here outlining the principal challenges that were faced, and some representative results for the hysteretic steel damping device that was tested.

The prototype building for the work is a single storey braced frame supporting a mass of 8.36 tonnes at a height of 3 m—a side elevation of the frame is shown in figure 5a. The frame is 3 m square in plan and is designed to be shaken in one direction—the frames in the other direction are heavily cross-braced. The damper is placed between the braces and the top beam. There is one frame on each side.

Two dampers were tested in the program: i) a yielding steel element invented by Prof Dorka at the University of Kassel and ii) Jarret® visco-elastic dampers. The devices were supplied by JRC Ispra.

The tests posed some interesting problems. Figure 5b shows a drawing of the Oxford set-up. At an early stage to ease the connection of the device it was decided to represent each brace by a separate 100 kN actuator. The damper is securely bolted onto a mounting plate (see Figure 6) which is bolted and dowelled to the concrete laboratory floor. The control system applied equal but opposite forces in the actuators meaning the primary force on the specimen was lateral, as in the prototype frame.

The Dorka element is manufactured from a short length of SHS (105 × 105 S355) and a single steel diaphragm is welded to close the section at the mid-point of the tube. The device is loaded transversely to its longitudinal axis of the tube which loads the diaphragm in shear. The relative displacement across the device was measured by an RDP® LVDT with a range of ±25 mm. The Jarret® damper has a very high initial stiffness but softens extensively at larger displacements. To capture these displacements a transducer with an appropriate range was required. The electrical noise on the transducer, 30 μm pk-pk, is not large but because of the high stiffness of the specimen did cause problems. The noise level on this transducer was significant at small displacements, and although the signal could be filtered to reduce it adequately, a significant phase lag was introduced.

4.1 *Displacement feedback algorithm—numerical model of the frame*

The original intention was to perform a displacement controlled test with the force as the feedback—the standard hybrid test loop. During the initial tests, however, it became apparent that this would not work. To produce a lateral load in the specimen the force in the second actuator was slaved to the first (in retrospect it would have been better to use a single actuator and a less realistic loading arrangement). The combination of filtering the displacement and slaving the second actuator introduced a significant phase lag into the applied lateral load. Stable displacement control could not be achieved.

The solution was to change the control variable to the force in the actuator, in this way the specimen displacement became the feedback signal as shown in Figure 7. The equivalent model for the numerical substructure is then given in Figure 8. The stiffness of the columns is represented by k_c and the damping rate by λ. The additional stiffness of the braces is k_b. The horizontal component of the brace force, F, is equal to the force in the specimen, which is also applied to the mass, m. The ground acceleration is $\ddot{x}_g$ and the displacement of the mass relative to the ground is x.

The state space equations for the numerical model are:

$$\begin{Bmatrix} \ddot{x} \\ \dot{x} \end{Bmatrix} = \begin{bmatrix} -2c\omega_b & -\omega_b^2 \\ 1 & 0 \end{bmatrix} \begin{Bmatrix} \dot{x} \\ x \end{Bmatrix} + \begin{bmatrix} \omega_b^2 - \omega_u^2 & -1 \\ 0 & 0 \end{bmatrix} \begin{Bmatrix} \delta \\ \ddot{x}_g \end{Bmatrix}$$

$$\begin{Bmatrix} F \\ x \end{Bmatrix} = \begin{bmatrix} 0 & -m(\omega_b^2 - \omega_u^2) \\ 0 & 1 \end{bmatrix} \begin{Bmatrix} \dot{x} \\ x \end{Bmatrix} + \begin{bmatrix} m(\omega_b^2 - \omega_u^2) & 0 \\ 0 & 0 \end{bmatrix} \begin{Bmatrix} \delta \\ \ddot{x}_g \end{Bmatrix}$$

where

$$\omega_b^2 = (k_c + k_b)/m$$

$$\omega_u^2 = k_c/m$$

and c is the damping coefficient of the unbraced frame. The numerical model can be completely defined in terms of the mass and the braced and unbraced natural frequencies of the frame with the

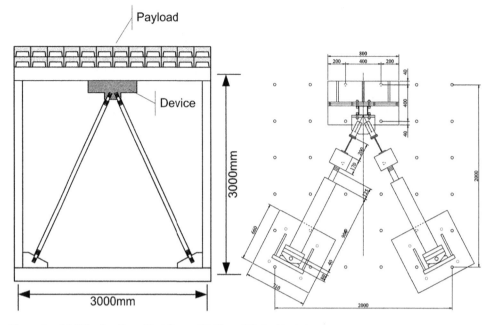

Figure 5.   (a) Side elevation of test frame, (b) Plan of Oxford test rig.

Figure 6.   View of the Dorka specimen in the testing rig.

damping factor of the free vibration. In tests at JRC Ispra on the prototype building the dynamic characteristics of the bare frame (without the bracing elements) were determined in sine sweep tests to be 2.60 Hz natural frequency with 3.02% damping. With the damper in place the natural frequency was 8.57 Hz with 5.3% damping.

Stable control was achieved by creating two control modes, one controlled the magnitude of the forces in both actuators and the other corrected the small differences between them. Each mode was controlled by an MCS controller. It was possible to filter the displacement feedback using a

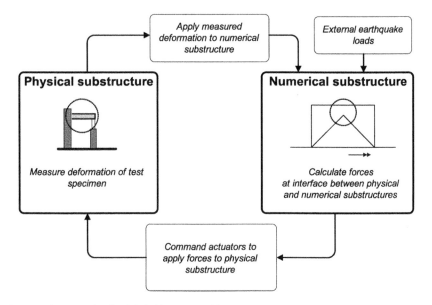

Figure 7.   Displacement feedback hybrid test control loop.

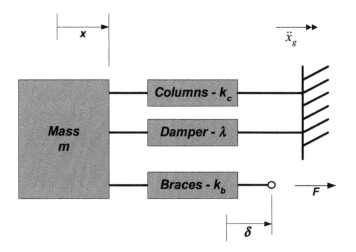

Figure 8.   Schematic diagram of displacement feedback numerical model.

low-pass 4-pole Butterworth filter with a pass band set at 70 Hz, which reduced the noise from the signal appreciably. Increasing the attenuation beyond this level introduced further phase lag, which caused stability problems.

4.2   *Control loops and tuning*

The first stage of setting up the control system was to adjust the inner Instron® actuator control gains to achieve a reasonable but not optimised response, since the MCS controllers would perform the final optimisation. The outer control loop was then switched in and the numerical model was driven by a square-wave ground acceleration. The system response was then optimised by alternately tuning the separate MCS controllers for each loop. Deciding when to stop tuning was a matter of judgment. If tuned for too long, the system went unstable and the tuning had to be restarted. As a final step, the earthquake was applied and the results checked. Further tuning was performed if it was thought that the results could be improved.

4.3 *Earthquake records and results*

Two earthquake inputs were used in the program, the El Centro record (20 s) and a 10 s record supplied by JRC Ispra and shown in Figure 9, which fitted the EC8 design spectrum. The El Centro record was used at 0.1 g and 0.2 g peak ground acceleration. The EC8 record, however, was designed to give a more balanced acceleration and displacement demand. This meant that greater magnifications could be used, and the range of acceleration was extended to 1.4 g for the EC8 tests.

Figure 10 shows the time history of the displacement across the device for the EC8 1.2 g pga test, which exhibited a large degree of plasticity, and the detail around the peak at 3 s is shown in Figure 11. The period lengthening at the major excursion can be clearly seen when this section is compared with the return to elastic oscillations following the peak at around 7.8 Hz, corresponding to the natural frequency of the model system. The force error is small and is mainly due to the 8 ms delay. The system faithfully reproduces the peaks and the profile of the force trace with, of course, the time delay.

Finally, the hysteresis curve for the specimen is given in Figure 12 for a linear test at 0.2 g and the 1.2 g test mentioned above. The width of the 'hysteresis' curve in the 0.2 g figure is entirely

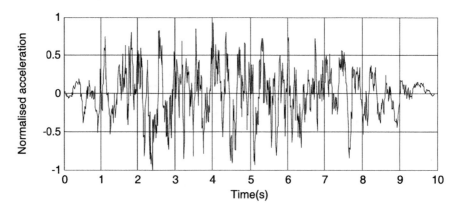

Figure 9.   Normalised EC8 earthquake record.

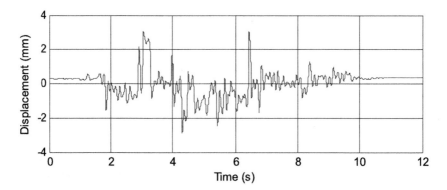

Figure 10.   Displacement on Dorka specimen EC8 PGA 1.2 g.

due to the effect of the noise. This is less significant in the later test which showed much greater plasticity. Similar results were obtained for all the Dorka specimen tests. These results show that the essential behaviour of the device is well reproduced in the tests, and that they can be considered a successful demonstration of the displacement feedback strategy.

5 CONCLUSIONS

Two test rigs have been described in outline. The first, a 3 DOF mass spring system was used as a test bed for comparing the various control strategies that have been proposed to enable real time substructure simulation. The results given show that real time substructuring can reproduce the behaviour of a two DOF mass system using a numerical model of one mass to drive the displacement of another. This was accurately achieved with this predominantly inertially excited system for natural frequencies of up to 10 Hz.

The second rig demonstrated that real time substructuring can also be successfully used to simulate a single degree of freedom system with a physical hysteretic spring element and numerical model of the remaining springs and the mass. Noise in the displacement signal combined with a very stiff specimen forced a change to the usual control loop but the tests were successfully preformed for the system which had a natural frequency of 8 Hz and 5% damping.

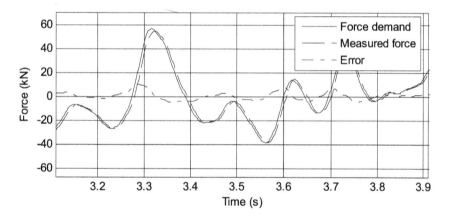

Figure 11. Detail of force variations on Dorka specimen EC8 PGA 1.2 g.

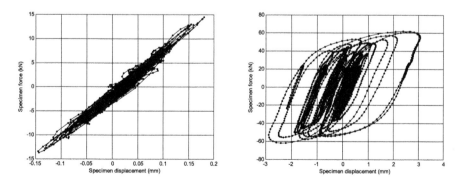

Figure 12. Dorka specimen force/displacement hysteresis curves for increasing levels of pga (EC8 record) left—0.2 g, right—1.2 g.

ACKNOWLEDGEMENTS

P.A. Bonnet was supported by the EPSRC under grant number GR/S03720/01. Other parts of the work presented in this paper is part of the NEFOREEE project (New Fields of Research in Earthquake Engineering Experimentation), funded by the European Commission (contract n. HPRI-CT-2001-50023) with the aim of the "Specific Research and Technological Development Programme of the Human Research Potential and the Socio-Economic Knowledge Base". The support of both these organisations is gratefully acknowledged.

REFERENCES

Blakeborough A., Williams M.S., Darby A.P. & Williams D.M. 2001. The development of real-time substructure testing. *Phil. Trans. Royal Soc. London*, A359, 1869–1891.

Bonnet P.A. 2006. The development of multi-axis real-time substructure testing. *D.Phil. thesis*, University of Oxford, Department of Engineering Science.

Bonnet P.A., Williams M.S. & Blakeborough A. 2005. Actuator performance in real time hybrid tests. *Proc. 1st Int. Conf. on Advances in Exp. Struct. Engng*, Nagoya, Japan.

Bonnet P.A., Williams M.S. & Blakeborough A. 2006. Compensation of actuator dynamics in real-time hybrid tests. *Proc. Instn Mech. Engrs, Part I: J. Systems & Control Engng*, 221, 251–264.

Bonnet P.A., Lim C.N., Williams, M.S., Blakeborough, A., Neild S.A., Stoten D.P. & Taylor C.A. 2007a. Real-time hybrid experiments with Newmark integration, MCSmd outer-loop control and multi-tasking strategies, *Earthquake Engng and Struct. Dyn.*, 36 119–141.

Bonnet P.A., Williams M.S. & Blakeborough A. 2007b. Evaluation of numerical time integration schemes for real-time hybrid testing *submitted to Earthquake Engng Struct. Dyn.*

Darby A.P., Williams M.S. & Blakeborough A. 2002. Stability and delay compensation for real-time substructure testing. *J. Engng Mech., ASCE*, 128(12), 1276–1284.

Horiuchi T., Nakagawa M., Sugano M. & Konno T. 1996. Development of real-time hybrid experimental system with actuator delay compensation. *Proc. 11th World Conf. on Earthquake Engng, Acapulco, Mexico*, Paper 660.

Molina F.J., Bairrao R., Blakeborough A., Bursi O., Tirelli D., Magonette G.E., Mouzakis H. & Williams M.S. 2006. Testing Performance Benchmark for Shaking Tables and Reaction Walls within the NEFOREEE Project, *First European Conference on Earthquake Engineering and Seismology Geneva, Switzerland, 3–8 September 2006* Paper Number: 303.

Stoten D.P. 1993. An overview of the minimal control synthesis algorithm. *Proc. I. Mech. E. Conf. on Aerospace Hydraulics and Systems, London*, Paper C474-033.

Wallace M.I., Sieber J., Neild S.A., Wagg D.J. & Krauskopf B. 2005a. Stability analysis of real-time dynamic substructuring using delay differential equation models. *Earthquake Engng Struct. Dyn.*, 34(15), 1817–1832.

Wallace M.I., Wagg D.J. & Neild S.A. 2005b. An adaptive polynomial based forward prediction algorithm for multi-actuator real-time dynamic substructuring, *Phil. Trans. Royal Soc. London* A461, 3807–3826.

Applications

CHAPTER 15

UI-SIMCOR: A global platform for hybrid distributed simulation

O.S. Kwon
*Post-doctoral research associate, Mid-America Earthquake Center, Department of Civil
and Environmental Engineering, University of Illinois at Urbana-Champaign, Urbana, IL, USA*

A.S. Elnashai
*Bill and Elaine Hall Endowed Professor, Department of Civil and Environmental Engineering,
Director, Mid-America Earthquake Center, University of Illinois at Urbana-Champaign, Urbana, IL, USA*

B.F. Spencer
*Nathan M. and Anne M. Newmark Endowed Chair,
Department of Civil and Environmental Engineering,
University of Illinois at Urbana-Champaign, Urbana, IL, USA*

ABSTRACT: Assessment of complex systems under extreme loading scenarios continues to be a formidable challenge that hampers development of effective mitigation, response and recovery measures. To overcome this difficulty, a new framework has been established at the University of Illinois, Urbana-Champaign, for combining seamlessly any number of testing sites with an unlimited number of analysis software packages in one single integrated hybrid (testing-analysis) distributed (different geographical locations) simulation of complex interacting systems. The framework, UI-SimCor, communicates with sites and programs through application program interfaces to integrate contributions from the various components of a complex system, such as buttresses and soil, or a bridge with abutments, piles and soil. This Chapter outlines the concept underlying UI-SimCor, its communication mechanisms and scope, and two application examples. The simulation coordinator UI-SimCor is freely available for use by researchers and practitioners and is effective in detailed and advanced assessment of complex systems under static and dynamic loading conditions.

1 INTRODUCTION

Analytically-oriented researchers have been developing applications to predict structural response based on principles of mechanics and/or observational-empirical data utilizing readily accessible computational resources. The ensuing analytical platforms are diverse in nature and have excellent problem-solving capabilities. Unfortunately, most or even all of these developments are limited to solving a specific set of relatively narrow problems of components within complex structural systems. An approach that has the minimum assumptions and provides the best available option is to model each component using the most suitable analytical model and integrating the various contributions into a fully interacting system. Whereas in theory the objective of accounting for interacting inelastic components could be achieved within one analysis platform, this possibility is not achievable with any existing package, and is unlikely to happen in the near future. It is indeed a fact that different analysis programs exhibit strengths and weaknesses and that combining programs with no restrictions placed on the selection is the obvious and only way forward.

Laboratory tests are one of the three fundamental sources of knowledge from which understanding of the behavior of structural systems can be attained; the other being field observations and analytical simulations. Due to the dimensions of civil engineering structures, such as buildings, bridges and utility networks, experiments are usually conducted on the most vulnerable components of a system and often at a reduced scale. Currently, the number of full scale complete structure tests is very limited. Examples of full scale system tests are Negro et al. (1996), Molina et al. (1999), Pinho & Elnashai (2000), Chen et al. (2003), and Jeong & Elnashai (2005). Even in the aforementioned cases, the foundations and soil were not modeled. A system by which a number of laboratories could combine their capabilities to undertake a set of integrated component tests of structural and geotechnical elements for example would provide an exceptionally attractive option for assessment of complex interacting systems with neither the assumptions necessary for conducting stable inelastic dynamic analysis, nor the limitations of small scale testing that would be required to fit all components into one laboratory.

The case is made above for distributed analysis, in contrast to using one analytical platform, and distributed testing, in contrast to using one experimental facility. There also exists a combination between the two, once the concept of a distributed representation is accepted. This 'Hybrid Simulation' approach has been subject to extensive research in recent years, (Watanabe et al., 1999, NSF, 2000, Tsai et al., 2003, Kwon et al., 2005, Pan et al., 2005, and Takahashi et al., 2006). It has hitherto remained, however, a rather arduous task that requires extensive knowledge of both experimental and analytical tools, their detailed input-output requirements, and necessitates considerable programming effort. The procedures have indeed not been sufficiently robust and had therefore remained in the advanced research domain, not in the persistent application domain.

This study addresses the above problem and proposes a simple, transparent and fully modular framework that allows the utilization of analytical platforms alongside experimental facilities for the integrated simulation of a large complex system. Whereas the framework presented is simple and intuitive, its impact on structural and geotechnical research is substantial. The approach utilizes pseudo-dynamic (PSD) simulation, distributed analysis and experimentation. It enables the combination of unique analysis applications in various fields and promotes collaboration of nationally and internationally distributed experimental and analytical simulation sites interested in large complex systems. The framework presented in this study is an extension of the previous development by Kwon et al. (2005). The following section provides brief conceptual background on the framework followed by the architecture and dataflow of the development. There have been several experimental and analytical applications of the developed framework. Among those, two representative experimental examples are introduced with summary of the potential of the framework.

2 THEORETICAL FORMULATION AND IMPLEMENTATION

The concept and architecture of a framework for multi-platform hybrid simulation is reviewed in this section. The developed framework is based on PSD simulation which has been conducted in many research institutes in the past. The conventional approaches, however, are limited to a specific experimental setup or to a specific analysis platform for which the PSD simulation is developed. The framework proposed n this Chapter allows the integration of various analysis platforms and experimental sites by employing widely adopted communication protocols, as well as a transparent and object-oriented programming architecture.

2.1 *Conceptual background*

The PSD test method has been investigated by researchers for more than thirty years. One of the earliest developments of the PSD method was by Hakuno et al. (1969) and Takanashi et al. (1975), which have evolved toward substructure (Dermitzakis & Mahin, 1985) and distributed PSD test (Watanabe et al., 2001). In these conventional PSD methods predicted displacements are imposed and measured restoring forces are used in the time integration scheme. These methods are in mature state in comparison with newly explored PSD test field such as real time testing (Nakashima et al., 1992 and Juan & Spencer, 2006), continuous PSD testing (Takanashi & Ohi, 1983), and effective force testing (Dimig et al., 1999). The proposed framework adopted conventional PSD testing scheme with its well established theory.

In a conventional PSD test, the structural mass, damping, and inertial forces are defined within a computational module. The predicted structural deformation at the control points is statically applied to a structure to estimate the restoring force vector. In a conventional PSD test of a whole structure, such as the three-storey frame depicted in Figure 1 (a), degrees of freedom (DOFs) with lumped masses are included in the equations of motion. If the three-story structure is pseudo-dynamically tested, it is assumed that the mass of each floor can be lumped at a single control point, and one actuator per story is normally used to apply inertial forces, for planar structures. Thus the computational module handles the equations of motion with three translational DOFs.

The experimental specimen for the PSD test may also be represented numerically, as shown in Figure 1 (b). The analytical model may use refined meshes to capture the initiation and propagation of damage. Hence the model may include a larger number of DOFs than the equations of motion where only the DOFs with lumped masses are used. The predicted displacements at the control points are applied, and the corresponding restoring forces are returned to the equations of motion. Where sub-structuring is required, force equilibrium and displacement compatibility should be satisfied at interfaces between substructured components. Hence, the control points should include nodes at interfaces, as well as nodes with lumped masses. The equations of motion subject to time integration should also include DOFs at lumped masses and interface nodes.

When testing a critical element and analyzing the rest of the structure, substructured PSD simulation should be used. In the conventional approaches for substructured PSD simulation, a single analysis platform is combined with a time integration module, as shown in Figure 1 (b). This approach is adequate if the adopted analysis platform can represent the true structural responses. In most situations, however, the analytical platform is limited to dealing with a simple nonlinear model. By completely separating the restoring force modules from the time integration scheme, Figure 1 (c), and by allowing an unhindered combination of restoring forces from various analytical modules, a complex structural system can be accurately modeled. In the proposed framework, the PSD test algorithm itself is identical to the conventional approach. But the way it combines several restoring force modules, whether analytical or experimental, and the communications between modules are the most distinctive characteristics of the development. The architecture of the framework, communication protocols, and simulation procedure are introduced in the following sections.

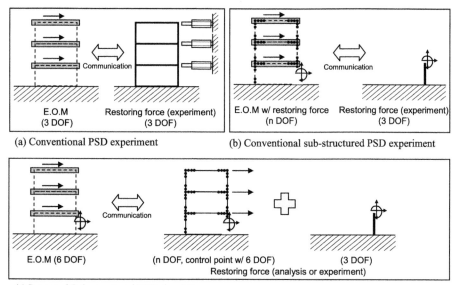

(a) Conventional PSD experiment

(b) Conventional sub-structured PSD experiment

(c) Proposed Sub-structured PSD simulation

Figure 1.   Substructuring of PSD simulation.

2.2   *System software architecture*

The basic concept of the framework is that analytical models associated with various platforms or experimental specimens are considered as super-elements with many DOFs. Each of these elements are solved on a single computer or on different computers connected through the network. Figure 2 illustrates the overall architecture of the framework, termed UI-SimCor. The main routine shown in the figure enforces equilibrium and conducts dynamic time integration. In this process, the structural model is fully encapsulated as objects of a class. Hence it is straightforward to add new time integration or methods to enforce static equilibrium.

The framework UI-SimCor is written in an object-oriented programming environment. There are two classes in UI-SimCor namely MDL_RF (restoring force module) and MDL_AUX (auxiliary module). The objects of MDL_RF class represent structural components. The main functionality of this class is abstraction of the structural components at remote sites. The main routines such as dynamic integration schemes impose displacements onto the structural components and retrieve restoring forces without consideration of communication with remote sites regardless of whether the components are experimental specimens or analytical models. This abstraction allows exceptionally easy implementation of new simulation tools and components.

Another important functionality of the MDL_RF class is communication. When the main analysis routines impose a displacement on a structural component represented by an object of MDL_RF class, the object reformats the data for the pre-specified protocol, opens connections to the remote sites, and sends the reformatted data. Six types of communication protocols are implemented in the current release. These are introduced in the following section. MDL_RF class includes other functionalities such as checking force and displacement capacities at every time step. In addition, the object of MDL_RF class shows the communication status and monitors communicated values at each time step. MDL_AUX class is used to control experimental hardware other than actuators. The object of this class has a function to send out pre-specified commands to remote sites.

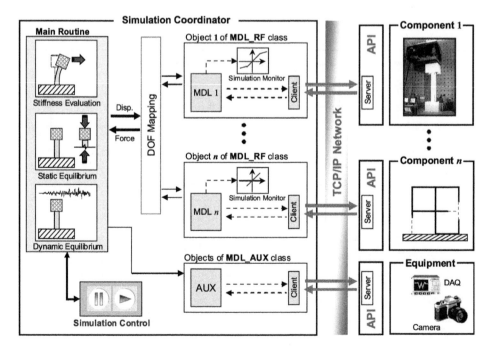

Figure 2.   Architecture of proposed framework.

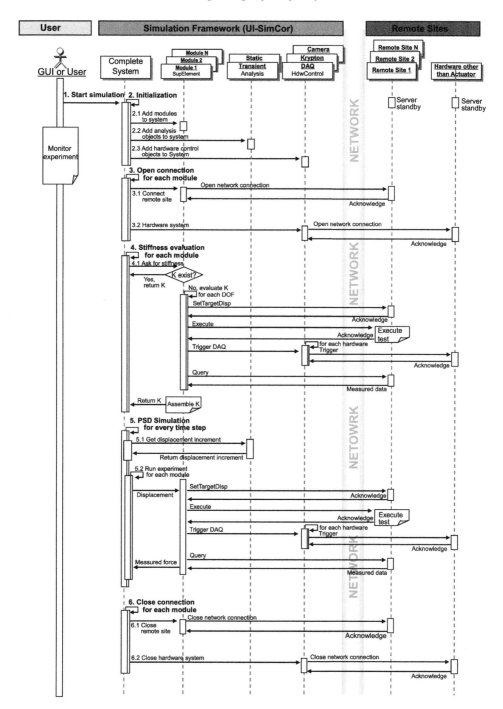

Figure 3.   Simulation procedure and data flow.

Upon reception of the command, the remote sites can take actions such as taking pictures or triggering data acquisition.

At remote sites, it is necessary to have an Application Program Interface (API) which open ports for connection from main framework, impose displacements to analytical model or experimental specimen, and send measured data. The APIs for analytical platforms have been developed for Zeus-NL (Elnashai et al., 2004), OpenSees (McKenna & Fenves, 2001), FedeasLab (Filippou & Constantinides, 2004), and ABAQUS (Hibbit et al., 2001). The API for VecTor2 (Vecchio & Wong, 2003) is under development.

2.3   *Simulation procedure and data flow*

A typical simulation procedure where three communication layers are labeled as 'User', 'Simulation Framework', and 'Remote Sites' is illustrated in Figure 3. The user of the hybrid simulation framework initiates the procedure, monitors its current status and pauses the simulation whenever necessary, based on warning messages. UI-SIMCOR is responsible for initialization, stiffness estimation, time integration, and communication with remote sites. The remote sites are responsible for running analysis or experiments under the commanded displacements and returning the measured resistance. The simulation procedure shown in Figure 3 is for a configuration with the Network for Earthquake Engineering Simulation (NEES) Telecommunication Control Protocol (NTCP) introduced in a subsequent section. The data flow shown in Figure 3 may vary depending on the protocols or simulation configuration used.

2.4   *Communication protocols*

The communication through the network following standard protocol is one of the most important requirements for geographically distributed hybrid simulations. In the proposed framework, six communication protocols are implemented: NTCP, LabView1, LabView2, TCP/IP, NEES Hybrid Simulation Communications Protocol (NHCP), and a protocol for OpenFresco (Takahashi & Fenves, 2006). To promote collaboration of equipment sites across the USA, NEES consortium has developed a standard communication protocol, NTCP (NEESgrid Teleoperation Control Protocol, Pearlman et al., 2004). NTCP allows secure communications between remote sites through the NTCP server. LabView1 and LabView2 protocols are communication protocol for which data are exchanged in ASCII format. The ASCII format data is very practical as all commands and values can be easily interpreted. But the format requires significant overhead as it needs to convert data from binary format to ASCII format at every simulation step. And also the converted data demand much larger network traffic. Thus in addition to these protocols, a binary format communication protocol, referred as TCP/IP in UI-SimCor, is also implemented. In the past few months, NEESit has been developing a NHCP protocol, a successor of NTCP. The earliest version of NHCP is also implemented in UI-SimCor. UI-SimCor can communicate with OpenFresco (Takahashi & Fenves, 2006) which provides a versatile interface to control experimental equipments. In addition to these already implemented communication protocols, any other protocols can be easily implemented. These versatilities in communication allow potential involvement of wide range of equipment sites and analysis platforms.

3   FRAMEWORK VALIDATION WITH THREE-SITE HYBRID SIMULATION

The main objective of the three-site hybrid simulation example described hereafter is to verify the proposed framework and check its compatibility with other experimental sites. Three sites are involved in this project. These are the University of Illinois at Urbana-Champaign (UIUC), the University of California at Berkeley (UCB), and San Diego Supercomputer Center (SDSC). Each experimental site is equipped with a small testing facility developed for the verification of a hybrid simulation; MiniMOST 1 (Gehrig, 2004) at UIUC and SDSC, μ-NEES (Schellenberg &

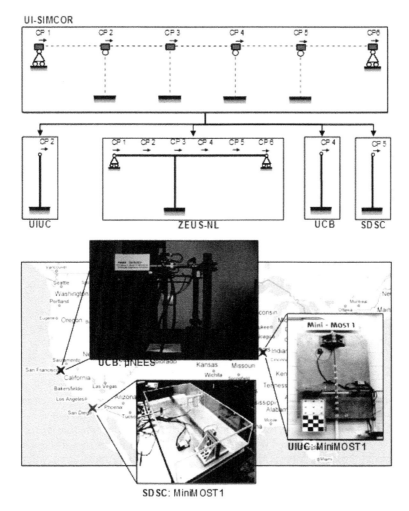

Figure 4.   Simulation configuration of three-site experiment.

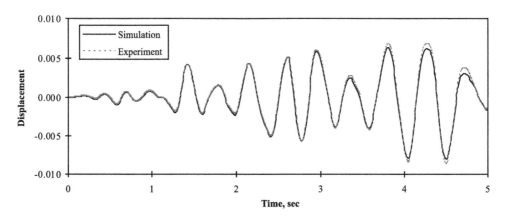

Figure 5.   Comparison of analytical and experimental results.

Mahin, 2006) at UCB. The MiniMOST 1 specimens behave in linear elastic range while the specimen in μ-NEES behaves fully in inelastic range. It is considered that the experimental specimens from three sites represent piers of a bridge. The remaining structural elements are modeled using Zeus-NL, Figure 4. The distributed simulation was carried out at a rate of 6.5 sec/step. The slow simulation rate was needed to allow for the stabilization of the load cell reading at UIUC and SDSC. Figure 5 compares the response from three-site experiment and PSD simulation with analytical model of experimental equipments. The experimental and analytical results are very close. The slight difference is caused by inaccurate representation of the inelastic behavior of the μ-NEES specimen with a hysteretic spring model. This project verified that the proposed framework runs reliably with minimum effort for customization at each of the participating remote site.

4   MULTI-SITE SOIL-STRUCTURE-FOUNDATION INTERACTION TEST

The main objective of MISST (Multi-Site Soil-Structure-Foundation Interaction, Spencer et al., 2006) project was to demonstrate the potential of NEES to investigate a structural-geotechnical system; a setup that has not been studied before. The tested bridge is based on the Collector-Distributor 36 of the I-10 Santa Monica Freeway that was severely damaged during the Northridge earthquake in January 1994. In this demonstration, two experimental sites (one pier at UIUC and another pier at Lehigh University, LU) and two analytical models (geotechnical model at Rensselaer Polytechnic Institute, RPI, and structural model at UIUC) are integrated using UI-SimCor, as shown in Figure 6. To satisfy capacity limitations of the laboratory equipment, a ½ scale model of prototype pier was constructed and tested at UIUC. The diameter of tested specimen was 24 inches with reinforcement ratio of 3.11% and 0.176% for longitudinal and transverse direction. Several hybrid simulations were carried out. These simulations included both small and large amplitude tests. The small amplitude test was intended to verify the functionality of all components and equipment while the large amplitude tests were intended to replicate the observed damage in the prototype structure. Two earthquake records that were captured during the Northridge earthquake of 1994 were employed during these simulations. The first record was strong motion data collected at the Santa Monica City Hall which had a peak ground acceleration (PGA) of

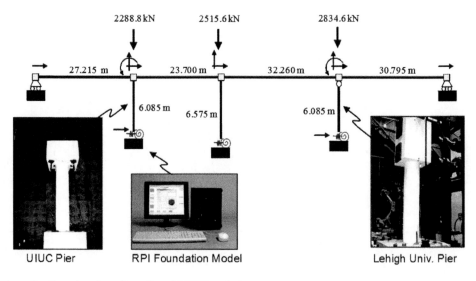

Figure 6.   Experiment configuration of MISST project.

0.37 g. The second record was collected at the Newhall Fire Station and had a PGA of 0.58 g. In both cases, the acceleration record was applied along the longitudinal direction of to the bridge structure.

The coordination and communication of the three sites, UIUC, Lehigh, and RPI, for the five component hybrid and geographically distributed simulation worked seamlessly. Despite the brittle nature fo the shear-critical piers, the simulation was able to continue well past the initial shear failures observed at both the UIUC and Lehigh sites. Furthermore, the redistribution of forces between the two sites with the bridge piers as either of the two suffered partial failure shows that full interaction was taking place between the distant sites. Thus the simulation system which includes all NEESgrid components, UI-SimCor, the analytical modules, and all experimental equipment and components at both UIUC and Lehigh proved to be quite effective and robust. Moreover, the failure modes obtained are similar to those in the prototype observed following the 1994 Northridge earthquake. Thus, the observed and complex field behavior of a complicated structural system was successfully reproduced. Not only does this create an opportunity to address or propose new design approaches for bridge structures, but also clearly demonstrates how NEES can be applied to address problems which have previously been unapproachable to the earthquake engineering community.

5   CONCLUSIONS

In this Chapter, a framework for multi-platform distributed earthquake simulation is introduced. The framework allows concurrent utilization of various analysis platforms and testing sites each of which is deemed the best environment for simulating a specific feature of the complex interacting system. For example, the geotechnical constitutive relationships implemented in OpenSees may be combined with the powerful fiber-based analysis capabilities of Zeus-NL to provide the most appropriate and accurate assessment procedure. The object-oriented program architecture allows extremely simple extension of the framework to new integration schemes, analysis methods, testing sites and analysis platforms. The framework has been verified through various hybrid experimental-analytical studies as well as analytical simulations. In this Chapter, two distinct hybrid simulations are introduced to demonstrate the potential of the proposed framework. One of the examples involves three experimental sites distributed across the USA. Each experimental specimen represents a bridge pier. The remainder of the bridge is represented by an analytical model in Zeus-NL, the MAE Center analysis platform. The hybrid simulation proved that the framework can be easily applied to multiple equipment sites. The other example involves large scale piers of the Santa Monica Bridge collector-distributor 36 under combined loading condition (shear-flexural-axial loads). The simulation coordinator UI-SimCor combines the responses of two large scale piers at the University of Illinois and Lehigh University with the analytical components. The developed framework opens up significant opportunities for potential collaborative research in analysis, experiments, and hybrid simulations. The proposed framework, UI-SimCor, is available for download from http://neesforge.nees.org/projects/simcor/.

ACKNOWLEDGMENTS

This research was supported by Mid-America Earthquake Center project EE-3, the NEES project at UIUC, and NEES Inc. The Mid-America Earthquake Center is an Engineering Research Center funded by the National Science Foundation under cooperative agreement reference EEC-9701785. NEES at UIUC is funded by the George E. Brown, Jr. Network for Earthquake Engineering Simulation (NEES) program of the National Science Foundation under award reference CMS-0217325 and NEESgird is funded by the same NSF Program, award reference CMS-0117853.

REFERENCES

Chen, C.-H., Lai, W.-C., Cordova, P., Deierlein, G.G., and Tsai, K.-C., 2003. Pseudo-Dynamic Test of Full-Scale Rcs Frame:Part 1—Design, Construction and Testing. *Proceedings of International Workshop on Steel and Concrete Composite Constructions*, Taipei, Taiwan, 107–118.

Dermitzakis, S.N., and Mahin, S.A., 1985. *Development of substructuring techniques for on-line computer controlled seismic performance testing.* Report UCB/EERC-85/04, Earthquake Engineering Research Center, University of California, Berkeley.

Dimig, J., Shield, C., French, C., Bailey, F., and Clark, A., 1999. Effective force testing: A method of seismic simulation for structural testing. *Journal of Structural Engineering.* 125(9): 1028–1037.

Elnashai, A.S., Papanikolaou, V. and Lee, D.H., 2004, *Zeus-NL — A System for Inelastic Analysis of Structures.* MAE Center CD-Release 04-01.

Filippou, F.C., and Constantinides, M., 2004. *FEDEASLab Getting Started Guide and Simulation Examples.* Technical Report NEESgrid-2004-22: www.nees-grid.org.

Gehrig, D. 2004. *Guide to the NEESgrid Reference Implementation.* NEESgrid TR-2004-04 Hakuno, M., Shicawara, M., and Hara, T. 1969. Dynamic destructive test of a cantilever beam, controlled by an analog-computer. *Transactions of Japan Society of Civil Engineering*, 171:1–9

Hibbit, Karlsson, and Sorensen. 2001. ABAQUS *theory manual.* Version 6.2.

Jeong, S., and Elnashai, A.S., 2005. Analytical Assessment of an Irregular RC Frame for Full-Scale 3D Pseudo-Dynamic Testing. Part I: Analytical Model Verification. *Journal of Earthquake Engineering*, 9(1), 95–128.

Juan, C., and Spencer, B.F., 2006. Real-Time Hybrid Testing Using Model-Based Delay Compensation. *Proceedings of* 4th *International Conference on Earthquake Engineering*, Taipei, Taiwan.

Kwon, O.S., Nakata, N., Elnashai, A.S., and Spencer, B., 2005. A Framework for Multi-Site Distributed Simulation and Application to Complex Structural Systems. *Journal of Earthquake Engineering*, 9(5), 741–753.

McKenna, F., and Fenves, G.L., 2001. *The OpenSees command language manual*, version 1.2. Pacific Earthquake Engineering Research Center, Univ. of California at Berkeley.

Molina, F.J., Verzeletti, G., Magonette, G., Buchet, P., and Geradin, M., 1999. Bi-Directional Pseudo-dynamic Test of a Full-Size Three-Storey Building. *Earthquake Engineering and Structural Dynamics*, 28, 1541–1566.

Nakashima, M., Kato, H., and Takaoka, E., 1992. Development of Real-Time Pseudo Dynamic Testing. *Earthquake Engineering and Structural Dynamics*, 21(1), 79–92.

National Science Foundation, 2000. *Network for earthquake engineering simulation (NEES): system integration*, program solicitation. Report NSC00-7, U.S.A.

Negro, P., Pinto, A.V., Verzeletti, G., and Magonette, G.E., 1996. PsD Test on Four-Story R/C Building Designed According to Eurocodes. *Journal of Structural Engineering*, 122(12), 1409–1471.

Pan, P., Tada, M., and Nakashima, M., 2005. Online hybrid test by internet linkage of distributed test-analysis domains. *Earthquake Engineering and Structural Dynamics*, 34, 1407–1425.

Pearlman, L., D'Arcy, M., Johnson, E., Kesselman, C., and Plaszczak, P., 2004. *NEESgrid Teleoperation Control Protocol (NTCP).* Technical Report NEESgrid-2004-23.

Pinho, R., and Elnashai, A.S., 2000. *Dynamic Collapse Testing of a Full-Scale Four Storey RC Frame.* ISET Journal of Earthquake Technology, 37(4), 143–163.

Schellenberg, A., and Mahin, S. 2006. Integration of Hybrid Simulation within the General-Purpose Computational Framework OpenSees. 100th *Anniversary Earthquake Conference Commemorating the 1906 San Francisco Earthquake.*

Spencer, Jr., B.F., Elnashai, A., Kuchma, D., Kim, S., Holub, C., and Nakata, N. 2006. Multi-*Site Soil-Structure-Foundation Interaction Test (MISST).* University of Illinois at Urbana-Champaign.

Takahashi, Y., and Fenves, G.L., 2006. Software framework for distributed experimental-computational simulation of structural systems. *Earthquake Engineering and Structural Dynamics*, 35, 267–291.

Takanashi, K., Udagawa, K., Seki, M., Okada, T., and Tanaka, H., 1975. Nonlinear earthquake response analysis of structures by a computer-actuator on-line system. *Bulletin of Earthquake Resistant Structure Research Centre*, No. 8, Institute of Industrial Science, University of Tokyo, Japan.

Takanashi, K., and Ohi, K., 1983. Earthquake response analysis of steel structures by rapid computer-actuator online system, (1) a progress report, trial system and dynamic response of steel beams. *Bull. Earthquake Resistant Struct. Research Center* (ERS), Inst. of Industrial Sci., Univ. of Tokyo, Tokyo, Japan, 16: 103–109.

Tsai, K., Hsieh, S., Yang, Y., Wang, K., Wang, S., Yeh, C., Cheng, W., Hsu, C., and Huang, S., 2003. *Network Platform for Structural Experiment and Analysis* (I). NCREE-03-021, National Center for Research on Earthquake Engineering, Taiwan.

Vecchio, F.J., and Wong, P., 2003. *VecTor2 and FormWorks Manual*.

Watanabe, E., Kitada, T., Kunitomo, S., and Nagata, K., 2001. Parallel pseudodynamic seismic loading test on elevated bridge system through the Internet. *The Eight East Asia-Pacific Conference on Structural Engineering and Construction*, Singapore.

Watanabe, E., Sugiura, K., Nagata, K., Yamaguchi, T., and Niwa, K., 1999. Multi-phase Interaction Testing System by Means of the Internet. *Proceedings of* 1st *International Conference on Advances in Structural Engineering and Mechanics*, Seoul, Korea, 43–54.

CHAPTER 16

Real-time hybrid simulation of a seismically excited structure with large-scale Magneto-Rheological fluid dampers

R. Christenson & Y.Z. Lin
University of Connecticut, Storrs, Connecticut, USA

ABSTRACT: Magneto-Rheological (MR) fluid dampers have been identified as a particularly promising type of semiactive control device for hazard mitigation in civil engineering structures. Experimental testing is important to verify the performance of MR fluid dampers for seismic protection. Real-time hybrid testing, where only the critical components of the system are physically tested while the rest of the structure is simulated, can provide a cost-effective means for large-scale testing of semiactive controlled structures. This paper describes the real-time hybrid simulation of multiple large-scale MR fluid dampers to verify semiactive control for seismic protection.

1 INTRODUCTION

Structural control shows great potential for hazard mitigation in civil structures. Within structural control there are three primary classes of control devices: passive, active and semiactive. Passive devices are non-controllable, require little to no power, and are inherently stable. Active devices are controllable offering more robust and significant response reduction, however, they require significant power to operate and are not inherently stable. Semiactive devices combine the positive attributes of passive and active control devices in that they are controllable in nature, require little energy to operate, and are inherently stable (Spencer and Sain, 1997). Semiactive control appears to be particularly promising in achieving increased performance over passive control strategies while addressing a number of the challenges facing active control strategies (Soong and Spencer, 2002). Semiactive control provides supplemental damping and stiffness to more efficiently dissipate and redistribute the energy due to dynamic loads, thereby increasing the safety and performance of the structure. Because they cannot inject mechanical energy into the controlled system, semiactive devices are inherently stable and well suited for application to structures with potential to behave nonlinearly. Additionally, the low power requirements of semiactive devices ensure that during extreme events, when external power may not be available, the semiactive device can continue to fully function using an alternate power source.

Various semiactive control devices have been proposed for structural control of civil engineering structures, however, the Magneto-Rheological (MR) fluid damper appears to be a particularly promising type of semiactive control device (Dyke et al., 1996; and Yang, 2001). In addition to the controllability, stability (in a bounded-input bounded-output sense), and low power requirements inherent to semiactive devices, MR fluid dampers with their large temperature operating range and relatively small device size have the added benefits of: producing large control forces at low velocities and with very little stiction; possessing a high dynamic range (the ratio between maximum force and minimum force at any given time); and having no moving parts, thus reducing maintenance concerns and increasing the response time (compared to conventional variable-orifice dampers).

Experimental testing is critical to verify and prove innovative concepts in structural control (Housner et al., 1994). Semiactive control technology has reached a state of maturity where experimental tests and full-scale implementations are needed to demonstrate the feasibility and advance the acceptance of these devices. Full-scale experimental verification of structural control is a challenging proposition. In particular, when applied to buildings that can sustain damage (e.g. exhibit nonlinear material behavior) during severe seismic events experimental testing can be cost prohibitive, require significant time to rebuild damaged structures or components, and raise issues of repeatability and consistency of the tests. Real-time hybrid testing, where only the critical components of the system are physically tested while the rest of the structure is simulated, can provide a cost effective and safe means to test full-scale time-dependant controlled structures during severe seismic events. The Network for Earthquake Engineering Simulation (NEES) Fast Hybrid Test (FHT) facility at the University of Colorado at Boulder (CU) (http://nees.colorado.edu/) has real-time hybrid simulation capabilities used to experimentally verify large-scale MR fluid dampers. The simulated component in these tests is a 3-story building structure subjected to suites of ground motions with nonlinear material behavior in the beam elements. The physical component of the experiment is a highly nonlinear and rate-dependant large-scale semiactive MR fluid damper *placed* between the stories of the simulated building. The added benefit of hybrid simulation is that it allows for various damper control strategies and a wide range of system parameter values to be examined through the numerical model without modifying the experimental (physical) setup.

This paper describes the real-time hybrid simulation of multiple semiactive control devices to verify large-scale semiactive control for seismic protection. First the simulated and physical components of the hybrid simulation experiment are identified. Next, the real-time hybrid simulation test method employed at CU to test the controllable dampers is introduced and the use of virtual coupling to provide performance and stability is discussed. Results from a series of real-time hybrid simulation tests are presented and compared to purely simulated predictions.

2   REAL-TIME HYBRID SIMULATION COMPONENTS

Real-time hybrid simulation provides the capability to isolate and physically test critical components of a semiactive controlled structure. The tests are conducted in real-time to fully capture any rate dependencies. Real-time hybrid simulation allows for hundreds of repeatable tests to be conducted to examine various control strategies and a range of seismic events in both an efficient and timely manner. In hybrid simulation the experiment is partitioned into simulated and physical components. For full-scale experimental verification of semiactive control, the components of interest are the semiactive control devices. As such, the simulated component is the seismically excited building with potential nonlinear beam-column connections. The physical component consists of multiple large-scale semiactive MR fluid dampers. A schematic of the concept for this particular hybrid simulation is depicted in Figure 1 and the simulated and physical components are described subsequently.

2.1 *Simulated component*

The simulated component of the hybrid simulation consists of a seismically excited finite element model representing the North-South moment resisting frame of the Los Angeles 3-story SAC structure (Ohtori et al., 2004). The building model is intended to represent a typical steel frame structure that receives damage during a large seismic event. The semiactive control devices will be added to this structure as a retrofit. The frame is depicted in Figure 2. The nonlinear in-plane finite element model is developed in OpenSees (http://opensees.berkeley.edu/index.php). The in-plane finite element model consists of 27 elements interconnecting 20 nodes. Each node has three degrees-of-freedom (DOFs), horizontal, vertical, and rotational. The column elements are modeled as elastic beam-column elements. The simply supported beam elements in the far right bay are modeled as truss elements. The remaining beams are modeled as beam with hinges elements where plasticity is concentrated over a 10% hinge length at the ends of the beam. The material in the beam with hinges elements is a bilinear uniaxial steel material (elastic—perfectly plastic) with a yield strength of 345 MPa (50 ksi). The floor system is assumed to add sufficient lateral stiffness to the floor such that the horizontal DOFs are fixed laterally to a single horizontal DOF on each story. The inherent damping in the structure is modeled as Rayleigh damping with 4.3% damping assumed in the first and third modes. The first three natural frequencies, corresponding to the first three lateral modes of the structure, are 0.99 Hz, 3.06 Hz and 5.83 Hz. Corresponding transfer functions for the interstory drift to a ground acceleration input are provided in Figure 2 to illustrate the participation of the higher modes in the structure's response.

2.2 Physical component

The physical component of the hybrid simulation is three large-scale MR fluid dampers distributed as a retrofit between the stories on each level of the building frame. In this fashion the displacement across the damper is directly proportional to the interstory drift of the structure. The three large-scale MR fluid dampers manufactured by Lord Corporation (http://www.lord.com/), are shown in Figure 3 along with a schematic of the components of these large-scale semiactive dampers. Each damper is 1.47 m (58 inches) in length, weighs approximately 2.734 kN (615 lbs), and has an available stroke of 584 mm (23 inches). The damper's accumulator can accommodate a temperature change in the fluid of 45°C (80°F). Each damper can provide control forces of approximately 200 kN (45 kip).

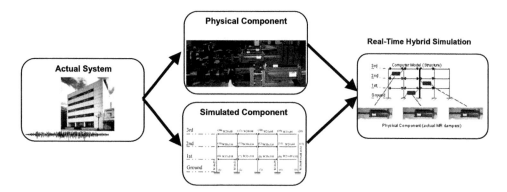

Figure 1.   Schematic depicting real-time hybrid simulation for a semiactive controlled seismically excited structure.

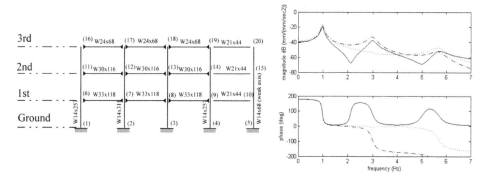

Figure 2.   Los Angeles 3-story SAC building model and corresponding transfer functions of interstory drift to ground acceleration.

(a)   (b)

Figure 3. Large-scale Magneto-Rheological (MR) fluid dampers: (a) picture during assembly; and (b) schematic of damper components.

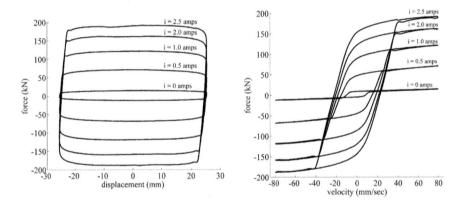

Figure 4. Typical behavior of a large-scale Magneto-Rheological (MR) fluid damper for various levels of constant current at a 25.4 mm amplitude 0.5 Hz sinusoidal excitation.

The dampers are controlled with a low voltage, current driven command signal. The coil resistance is approximately 4.8 Ω, while the inductance is approximately 5 H at 1 amp and 3 H at 2 amps. An Advanced Motion Controls PWM Servo-Amplifier (30A8DDE) is powered by a 70 volt DC, 5 amp unregulated linear power supply. The servo-amplifier is used to provide the command signal that controls the electromagnetic field for each damper. The PWM Servo-Amplifier is controlled by a 0–5 volt DC signal and utilizes pulse width modulation for current control. The input control signal can be switched at a rate of up to 1 kHz, although the rise time of the current signal is limited by the inductance of the MR damper to approximately 30 msec. Each damper has been fitted with a 1.5KE75A transient voltage suppressor to protect the MR damper electromagnetic coils from unintended and damaging voltage peaks.

To observe the controllable nature of the MR damper, a series of force versus displacement plots (hysteretic loops) and force versus velocity plots are shown in Figure 4 for incrementally increasing current command signal. From the plots it is observed that the damper force can be varied from 15 kN to 200 kN by providing a 0 to 2.5 amp command signal to the MR damper.

3   REAL-TIME HYBRID SIMULATION OF MR FLUID DAMPERS

Real-time hybrid simulation is conducted at the University of Colorado at Boulder (CU Boulder) Fast Hybrid Test (FHT) facility using a high-performance servo-hydraulic test system, modular

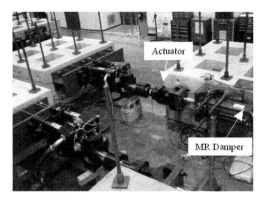

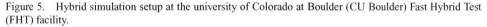

Figure 5.   Hybrid simulation setup at the university of Colorado at Boulder (CU Boulder) Fast Hybrid Test (FHT) facility.

reaction walls, modern digital control technologies, and the state-of-the-art hybrid simulation techniques. A picture of the experimental setup for the large-scale MR fluid dampers is shown in Figure 5. Real-time hybrid simulation, also called real-time dynamic substructuring, is a relatively new method of testing (Nakashima, 1999) that has its origins in pseudodynamic testing.

Pseudo-dynamic testing (PsD) has been traditionally used in earthquake engineering to test large scale critical componets (Mahin et al., 1989). The conventional PsD methods are not appropriate for structures requiring physical testing of rate-sensitive components. Various continuous PsD methods have been proposed to allow testing of strain-sensitive material (Magonette, 2001 and Mosqueda, 2006). Furthermore, hard real-time hybrid simulation has been proposed to fully capture strain, damping and inertial effects (Nakashima, 1999; Dimig, 1999; Horiuchi et al., 1999; Darby, 1999; Blakeborough et al., 2001; and Williams and Blakeborough, 2001). In real-time hybrid simulation the dynamics of the testing system and numerical integration scheme are critical (Horiuchi et al., 1999; Nakashima, 2001; and Zhao et al., 2003; Mercan and Ricles, 2007). The CU Boulder facility contains a state-of-the-art computer control system to coordinate the real-time hybrid testing, measuring and enforcing the essential continuity and equilibrium conditions between the physical and simulated components at each time step using high-speed hydraulic actuators. As the rate of loading is much faster than traditional hybrid testing, the numerical integration becomes a critical component.

The time integration scheme employed by the CU FHT system is the α-method, based on the implicit Hilbert-Hughes-Taylor α-method (Wei, 2005). The customized unconditionally stable implicit integration scheme allows CU Boulder FHT facility to achieve hard real-time performance. During the test the actuators are in continuous motion and the FHT algorithm ensures system equilibrium at each time step. The α-method uses a fixed number of 10 iterations to converge to equilibrium at each time step. The time step is 9.7656 msec, while each iteration is calculated at a hard 0.9766 msec (1024 Hz). The α-method is further modified to accommodate the anticipated nonlinear damping effect of the MR fluid dampers (Stauffer, 2006). In this approach the second order ordinary differential equation representing the discrete governing equations of motion of the system is given as

$$M\ddot{x} + s(\dot{x}) + r(x) = f \tag{1}$$

where M is the mass matrix, s is the damping force vector which is assumed to be a nonlinear function of the velocity, r is the restoring force which is a function of displacement, f is the vector of applied forces—which could be an equivalent force vector for seismic excitation, and x is the displacement vector of the system where a dot indicates a derivative with respect to time. Using a fixed number of Newton iterations it is shown (Stauffer, 2006) that the displacement can be repeatedly solved where for the kth Newton iteration of the nth time step is

$$x_{n+1}^{k+1} = x_{n+1}^k - \left[M + c_1 \left(\frac{\alpha}{\Delta t j \beta} C_i \right) \right]^{-1} f(x_{n+1}^k) \tag{2}$$

where $c_1 = \Delta t \beta (1 + \alpha)$, Δt is the time step, α and β are the Newmark parameters that define the variation of acceleration over a time step, C_i is the tangent damping—an estimate of the damping coefficient, and $f(x_{n+1}^k)$ is the external applied force vector at x_{n+1}^k. This new integration scheme is shown to provide improved accuracy, with respect to the prior stiffness-based scheme (Wei, 2005), when applied to the MR damper physical specimens.

The selection of C_i in this modified integration scheme is critical to the overall performance and accuracy of the hybrid simulation. Overestimating this initial damping estimate adds significant numerical damping into the simulation. This added damping can distort the performance of the supplemental damping devices in the hybrid simulation. Alternatively, selecting a C_i too small can result in unstable behavior observed in system chattering and instability as discussed subsequently.

In real-time hybrid simulation the dynamic interaction of the hydraulic actuator and physical component contains a velocity feedback of the hydraulic supply which can significantly effect the performance of the actuator system (Dyke et al., 1995). This behavior appears as an apparent time *delay* and is manifested in steady state or unstable high frequency oscillations (Kyrychko et al., 2006). The traditional approach to resolve the systematic error in real-time hybrid simulation due to test system dynamics is to reduce the actuator time *delay*. A variety of compensation methods have been proposed where the actuator command signal is extrapolated forward so that the time *delay* is minimized. Numerous researchers have employed nth-order polynomials (Horiuchi et al., 1999; Darby et al., 2001; Horiuchi and Konno, 2001; and Wu et al., 2006), while others have proposed model-based predictions (Wei, 2005, Carrion and Spencer, 2006) and other feedforward circuits and predictors (e.g. a Smith prediction controller (Smith, 1959)) in the actuator control loop (Magonette, 2001 and Arioui et al., 2003). Over predicting this compensation will add numerical damping into the hybrid simulation, again potentially distorting the results of the system. Under compensation may cause instability of the closed-loop system (Darby et al., 2002). These methods often assume the time *delay* resulting from the hydraulic actuators is constant. It is observed in initial testing at CU Boulder that the time delay for an experiment where the damper current is varied between 0 and 2.5 amps varies from 3 msec to 6 msec. To account for varying time delay Wallace et al. (2005) have proposed an adaptive predictive method.

An alternative approach employed at the University of Colorado at Boulder for the MR real-time hybrid simulation tests to improve system stability in the presence of a time delay is to introduce a virtual coupling between the physical and simulated components (Colgate et al., 1995; Adams and Hannaford, 1998, 1999; Zilles and Salisbury, 1995 and Ruspini et al., 1997). One realization of virtual coupling for the stability of the fast hybrid simulation with inherent time delay is to place a virtual spring and viscous damper in parallel between the physical and simulated components. This approach, like many of the compensation methods, is independent of a particular integration scheme and experimental system.

There is a tradeoff between performance and stability when employing virtual coupling. A large virtual stiffness, relative to the physical stiffness, corresponds to a high performance system. However, a large virtual stiffness does not help to mitigate the instabilities resulting from the actuator time *delay*. Reducing the virtual stiffness does increase the hybrid simulation stability (counterbalancing the actuator time *delay*), however also reduces the system's ability to impart the desired displacements of the simulated component onto the physical specimen by introducing compliance in this interface. The difference between the simulated and physical displacement introduces error into the simulation and represents decreased performance. The virtual coupling (as a ratio to the virtual stiffness) determines the frequency bandwidth of the virtual coupling.

The attractive feature in virtual coupling is that the inherent tradeoff between performance and stability adjusts according to the relative force of the physical component. When the physical restoring force is small, the hybrid simulation stability is good and the virtual stiffness can be selected to provide sufficient performance. As the physical restoring force becomes more significant relative to the simulated component (and relative to the fixed virtual stiffness) the simulation tends to exhibit more instability. However, the virtual stiffness in this case is relatively smaller relative to the physical stiffness and the virtual coupling provides increased stability traded for performance. Thus the virtual coupling is observed to provide a mechanism to deliver an appropriate balance of performance versus stability. This mechanism is desirable when considering physical specimens with large dynamic ranges and highly nonlinear behavior, such as the MR fluid dampers and is the reason virtual coupling is employed in these tests.

4   MR DAMPER REAL-TIME HYBRID SIMULATION RESULTS

The real-time hybrid simulation of multiple large-scale semiactive control devices distributed throughout a building frame is conducted to verify the seismic performance of the proposed system. Presented here are real-time hybrid simulations conducted for three physical MR fluid dampers, one for each story of the simulated semiactive controlled building. The measured damper force is multiplied by a factor of two before being applied to the structure. This procedure simulates multiple dampers at each level and increases the overall system performance.

The tangent damping, C_i, of Eq (2) is set to 100 kip-sec/in to provide a sufficient level of numerical convergence. The virtual coupling link parameters are set to provide a balance of performance and stability. The virtual stiffness is set to 4,000 kip/in. The virtual damping is set to

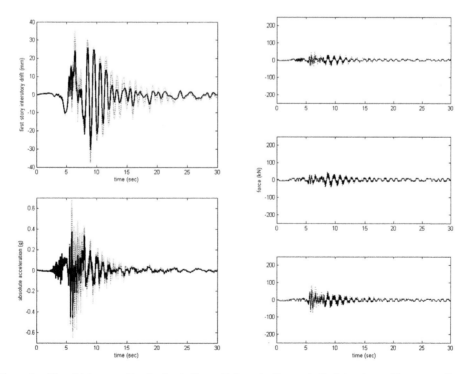

Figure 6.   Time history results of seismically excited semiactive controlled structure with constant 0 amp damper current. (black—hybrid simulation; grey dashed—Matlab simulation; solid grey—uncontrolled).

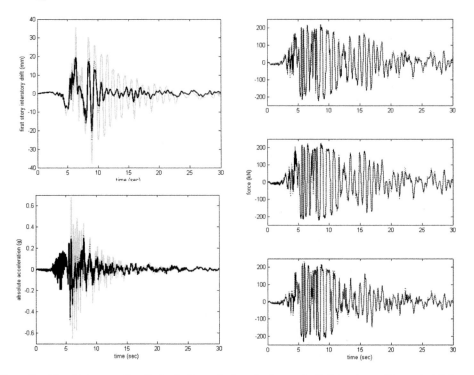

Figure 7.   Time history results of seismically excited semiactive controlled structure with constant 2.5 amp damper current. (black—hybrid simulation; grey dashed—Matlab simulation; solid grey—uncontrolled).

10 kip-sec/in. The displacement error of the hydraulic actuator introduced by the virtual coupling link is approximately 1/3 of the tracking error of the actuator. The virtual coupling link also adds sufficient stability to the system, reducing a steady state oscillation observed in the damper force from a 125 kN 14 Hz oscillation to a 4 kN, 8 Hz oscillation. Thus the virtual coupling link is noted to provide a sufficient level of stability.

A Matlab model is developed to simulate the entire system numerically. The Matlab building model is adapted from Ohtori et al. (2004). The Matlab MR damper model is a hyperbolic tangent model described in detail in Bass and Christenson (2006). The purely simulated system provides a baseline prediction to compare the real-time hybrid simulations in order to prove the accuracy of the predictions regarding the behavior of the MR fluid dampers.

The semiactive controlled system is subjected to the 1989 Loma Prieta earthquake (LA23), scaled to a factor of 0.82 for the SAC ground motion project to have a 2% probability of exceedence in 50 years. Three real-time hybrid simulations are presented. First, the dampers are operated in the passive off mode where 0 amps are sent to each damper. Next, the dampers are operated in the passive on mode, where 2.5 amps are sent to each damper. Lastly, a control signal determined from a clipped optimal H2/LQG control design and associated bang-bang controller (Christenson and Emmons, 2005) is sent to the dampers. The first story interstory drift ratio (the most critical interstory drift), the top story absolute acceleration (the largest story acceleration), and all three MR fluid damper forces are provided in Figures 7–9.

The first floor peak interstory drift of the MR damper controlled structure is reduced in the real-time hybrid simulations 20%, 46% and 44% beyond the uncontrolled peak interstory drift for passive off, passive on and active (H2/LQG and bang-bang) damper control, respectively. The top floor maximum absolute acceleration of the MR damper controlled structure is reduced in the real-time hybrid simulations 33%, 34% and 37% beyond the uncontrolled peak interstory drift for passive off, passive on and active damper control, respectively.

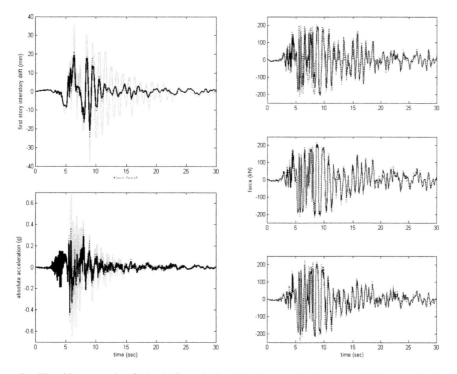

Figure 8.   Time history results of seismically excited semiactive controlled structure with time varying damper current. (black—hybrid simulation; grey dashed—Matlab simulation; solid grey—uncontrolled).

Table 1.   Comparison of real-time hybrid and purely simulated building responses for the SAC 1989 Loma Prieta earthquake (LA23).

Response	Passive off		Passive on		Active		
	Hybrid	Simulated	Hybrid	Simulated	Hybrid	Simulated	Uncontrolled
Peak drift ratio	1.02%	1.21%	0.64%	0.76%	0.68%	0.76%	1.23%
Permanent drift ratio	0%	0%	0%	0%	0%	0%	0.06%
Peak abs. accel. (g)	0.46	0.59	0.45	0.48	0.43	0.46	0.68
RMS abs. accel. (g)	0.07	0.09	0.05	0.05	0.05	0.05	0.10
# Plastic hinges	0	12	0	0	0	0	12
Energy dissip. structure	1.4	13.3	0.0	0.0	0.0	0.0	20.7

The responses for interstory drift, absolute story acceleration and damper force correspond well with the real-time hybrid simulation results, as given in Table 1. It is observed that for the passive on and actively controlled tests, with larger damper force, the real-time hybrid and pure simulations provide comparable results. The comparison of purely simulated and real-time hybrid simulation responses proves the accuracy of prior predictions regarding the behavior of the MR fluid dampers, a goal of hybrid simulation (Nakashima, 1999).

The results indicate that large-scale MR fluid dampers placed throughout a structure can provide significant response reduction and increased seismic protection to an existing steel frame structure. The results also document that full-scale experimental verification of semiactive control strategies for buildings exhibiting nonlinear behavior during large seismic events is accomplished through fast hybrid testing at the NEES FHT facility at CU Boulder.

5   CONCLUSIONS

Real-time hybrid simulation is successfully conducted at the CU Boulder NEES FHT facility. The hybrid tests physically tested three large-scale MR fluid dampers while simulating the seismic response of a 3-story steel frame structure. By conducting this type of test, the performance of the MR fluid dampers are experimentally verified as applied to a structure allowed to yield under severe dynamic loading. The real-time hybrid simulations provided a cost-effective means for large-scale testing of the critical components in a semiactive controlled structure.

The semiactive controlled system is subjected to the 1989 Loma Prieta earthquake (LA23), scaled to a factor of 0.82 for the SAC ground motion project. The first floor peak interstory drift of the MR damper controlled structure is reduced in the real-time hybrid simulations 20%, 46% and 44% beyond the uncontrolled peak interstory drift for passive off, passive on and active (H2/LQG and bang-bang) damper control, respectively. The top floor maximum absolute acceleration of the MR damper controlled structure is reduced in the real-time hybrid simulations 33%, 34% and 37% beyond the uncontrolled peak interstory drift for passive off, passive on and active damper control, respectively.

A comparison of purely simulated and real-time hybrid simulation responses illustrates the accuracy of prior predictions regarding the behavior of the MR fluid dampers. The results indicate that large-scale MR fluid dampers placed throughout a structure can provide significant response reduction and increased seismic protection to an existing steel frame structure. The results also document that full-scale experimental verification of semiactive control strategies for buildings exhibiting nonlinear behavior during large seismic events is accomplished through fast hybrid testing at the NEES FHT facility at CU.

ACKNOWLEDGMENTS

The writers gratefully acknowledge the support of this research by the National Science Foundation under Grant No. CMS-0612661 (pre-NEES Research) and Grant No. CMS-0402490 (NEES Operations). They would also like to thank the Lord Corporation for their generous support of this research. Finally, the writers would like to acknowledge the diligent efforts of the staff at the NEES FHT system at CU in helping to set up and conduct these experiments.

REFERENCES

Adams, Richard J., Moreyra, Manuel R. & Hannaford, Blake. 1998. Stability and performance of Haptic displays: Theory and experiments. *Proceedings ASME International Mechanical Engineering Congress and Exhibition*: 227–234.
Adams, Richard J. & Hannaford, Blake. 1999. Stable Haptic interaction with virtual environments. *IEEE Transactions on Robotics and Automation* 15(3): 465–474.
Arioui, H., Kheddar, A. & Mammar, S. 2003. A model-based controller for interactive delayed Haptic feedback virtual environments. *Journal of Intelligent and Robotic Systems* 37: 193–207.
Bass, B.J. & Christenson, R.E. 2007. System identification of a 200 kN Magneto-Rheological fluid damper for structural control in large-scale smart structures. *Proceedings of the American Control Conference*: New York City, NY.
Blakeborough, A., Williams, M.S., Darby, A.P. & Williams, D.M. 2001. The development of real-time substructure testing. *Philosophical Transactions: Mathematical, Physical and Engineering Sciences* 359(1786): 1869–1891.
Carrion, Juan E. & Spencer, B.F., Jr. 2006. Real-time hybrid testing using model-based delay compensation. *4th International Conference on Earthquake Engineering* paper No.299, Taipei, Taiwan.
Christenson, R.E., & Emmons, A.T. 2005. Semiactive structural control of a nonlinear building model: Considering reliability. *ASCE Structures Congress 2005*: Session on *Semiactive Control of Civil Structures*.
Colgate, J. Edward, Stanley, Michael C. & Brown, J. Michael. 1995. Issues in the haptic display of tool use. *IROS*.

Darby, A.P., Blakeborough, A. & Williams, M.S. 1999. Real-time substructure tests using hydraulic actuator. *Journal of Engineering Mechanics* 125(10): 1133–1139.

Darby, A.P., Blakeborough, A. & Williams, M. S. 2001. Improved control algorithm for real-time substructure testing. *Earthquake Engineering and Structural Dynamics* 30: 431–448.

Darby, A.P., Williams, M.S. & Blakeborough, A. 2002. Stability and delay compensation for real-time substructure testing. *Journal of Engineering Mechanics* 128(12): 1276–1284.

Dimig, J., Shield, C., French, C., Bailey, F. & Clark, A. 1999. Effective force testing: A method of seismic simulation for structural testing. *Journal of Structural Engineering* 125(9): 1028–1037.

Dyke, S.J., Spencer, B.F., Jr., Quast, P. & Sain, M.K. 1995. The role of control-structure interaction in protective system design. *Journal of Engineering Mechanics* 121(2): 322–338.

Dyke, S.J., Spencer, B.F., Jr., Sain, M.K. & Carlson, J.D. 1996. Modeling and control of Magnetorheological dampers for seismic response reduction. *Smart Materials and Structures* 5: 565–575.

Horiuchi, T., Inoue, M., Konno, T. & Namita, Y. 1999. Real-time hybrid experimental system with actuator delay compensation and its application to a piping system with energy absorber. *Earthquake Engineering and Structural Dynamics* 28(10): 1121–1141.

Horiuchi, T., Inoue, M., Konno, T. & Yamagishi, W. 1999. Development of a real-time hybrid experimental system using a shaking table (proposal of experimental concept and feasibility study with rigid secondary system). JSME International Journal 42(2): 255–264.

Horiuchi, Toshihiko & Konno, Takao. 2001. A new method for compensating actuator delay in real-time hybrid experiments. *Philosophical Transactions: Mathematical, Physical and Engineering Sciences* 359(1786): 1893–1909.

Housner, G.W., Soong, T.T. & Masri, S.F. 1994. Second generation of active structural control in civil engineering. *Proceedings 1st World Conference on Structural Control*: Pasadena: California, Panel: 3–18.

Kyrychko, Y.N., Blyuss, K.B., Gonzalez-Buelga, A., Hogan, S.J. & Wagg, D.J. 2006. Real-time dynamic substructuring in a coupled oscillator-pendulum system. *Proceedings of the Royal Society* 462: 1271–1294.

Magonette, Georges. 2001. Development and application of large-scale continuous pseudo-dynamic testing techniques. *Philosophical Transactions: Mathematical, Physical and Engineering Sciences* 359: 1771–1799.

Mahin, Stephen A., Shing, Pui-Shum, Thewalt, Christopher R.& Hanson, Robert D. 1989. Pseudodynamic test method-Current status and future directions. Journal of Structural Engineering 115(8): 2113–2128.

Mercan, Oya & Ricles, James M. 2007. Stability and accuracy analysis of outer loop dynamics in real-time pseudodynamic testing of SDOF systems. *Earthquake Engineering and Structural Dynamics* (in press).

Mosqueda, Gilberto, Stojadinovic, Bozidar, Hanley, Jason, Sivaselvan, Mettupalayam & Reinhorn, Andrei. 2006. Fast Hybrid Simulation with Geographically Distributed Substructures. *Proceedings of the 17th Analysis and Computation Specialty Conference*, St. Louis, Missouri.

Nakashima, Masayoshi & Masaoka, Nobuaki. 1999. Real-time on-line test for MDOF systems. *Earthquake Engineering and Structural Dynamics* 28: 393–420.

Nakashima, Masayoshi. 2001. Development, potential, and limitations of real-time online (pseudo-dynamic) testing. *Philosophical Transactions: Mathematical, Physical and Engineering Sciences* 359(1786): 1851–1867.

Ohtori, Y., Christenson, R.E., Spencer, B.F., Jr. Dyke, S.J. 2004. "Nonlinear benchmark control problem for seismically excited buildings." *ASCE Journal of Engineering Mechanics* 130(4): 366–385.

Ruspini, D.C., Kolarov, K. & Khatib, O. 1997. The Haptic display of complex graphical environments, in the *Proceedings of Computer Graphics SIGGRAPH '97*, Los Angeles, CA, 345–352.

Shing, P.B. & Mahin, S.A. 1987. Cumulative error effects in pseudodynamic tests. *Earthquake Engineering and Structural Dynamics* 15: 409–424.

Smith, O. 1959. A controller to overcome dead time. *ISA Journal* 6(2): 28–33.

Soong, T.T. & Spencer, B.F., Jr. 2002. Supplemental energy dissipation: State-of-the-Art and State-of-the-Practice. *Engineering Structures* 24: 243–259.

Spencer, B.F., Jr. & Sain, M.K. 1997. Controlling buildings: A new frontier in feedback," *IEEE Control Systems Magazine* 17: 19–35.

Stauffer, E. 2006. The CU-Boulder Fast Hybrid Test: Integration schemes for Fast Hybrid Testing. *Internal report*: Department of Civil Environmental and Architectural Engineering, University of Colorado at Boulder, CU-NEES-06-8, 5 pages.

Wallace, M.I., Sieber, J., Neild, S.A., Wagg, D.J. & Krauskopf, B. 2005. Stability analysis of real-time dynamic substructuring using delay differential equation models. *Earthquake Engineering and Structural Dynamics* 34: 1817–1832.

Wei, Zhong. 2005. Fast hybrid test system for substructure evaluation. *Ph.D dissertation*: University of Colorado, Boulder.

Williams, M.S. & Blakeborough, A. 2001. Laboratory testing of structures under dynamic loads: an introductory review. *Philosophical Transactions: Mathematical, Physical and Engineering Sciences* 359: 1651–1669.

Wu, B., Xu, G., Wang, Q. & Williams, M.S. 2006. Operator-splitting method for real-time substructure testing. *Earthquake Engineering and Structural Dynamics* 35: 293–314.

Yang. G. 2001. Large-scale Magnetorheological fluid damper for vibration mitigation: Modeling, testing and control. *Ph.D dissertation*: University of Notre Dame.

Zhao, J., French, C., Shield, C. & Posbergh, T. 2003. Considerations for the development of real-time dynamic testing using servo-hydraulic actuation. *Earthquake Engineering and Structural Dynamics* 32: 1773–1794.

Zilles, C.B. & Salisbury, J.K. 1995. A constraint-based God-object method for haptic display, *Proceedings of IEEE/RSJ International Conf. Intel Robots Systems*, Pittsburgh, PA, 146–151.

CHAPTER 17

Hybrid simulation at CRIEPI: Applications to soil structure interaction

K. Ohtomo & M. Sakai
Central Research Institute of Electric Power Industry, Chiba, Japan

Y. Dozono & M. Fukuyama
Mechanical Engineering Research Laboratory, Hitachi Ltd., Ibaraki, Japan

ABSTRACT: An advanced substructure hybrid test system is developed so that soil structure interaction problem can be satisfactorily evaluated. For this purpose, a nonlinear finite element analysis is introduced for the numerical part of the system. Direct time integral algorithm and numerical representation of a linear stiffness matrix are devised for stably implementing nonlinear analysis. The validity of the advanced hybrid test system is verified with a large-scale shake table test results on inelastic response of an underground reinforced concrete culvert. In addition, one of the directions to enhance the developed system is addressed.

1 INTRODUCTION

Soil structure interaction accompanied by material and/or geometric nonlinearity during a severe ground shaking is a great concern in earthquake resistance capability challenges on civil engineering structures. Subway tunnels were severely damaged during the 1995 Hyogoken Nanbu earthquake due to excessive lateral force resulting from a soil structure interaction (Iida et al., 1996). Matsui et al. (2004) studied on seismic performance of in-ground reinforced structure based on a large shale table test and validated a nonlinear finite element analysis for evaluating inelastic deformation of an underground structure. As is reported extensively, the deeply embedded structures of Kashiwazaki-Kariwa nuclear plants, Niigata, Japan showed interaction with soil. Many of the problems on non-safety related items were also induced by ground deformation (IAEA, 2007). This emphasizes the effects of soil structure interaction on seismic behavior of structures are still crucial for an earthquake resistant design practice.

A substructure hybrid experiment can substantially cover seismic performance study on versatile civil engineering and building structures (Takanashi & Nakashima, 1987; Mahin et al., 1989). One of such a promising application is a dynamic soil structure interaction problem. However, current substructure hybrid tests usually employ a lumped mass model for a numerical analysis with relatively smaller degrees of freedom. Use of a nonlinear finite element analysis will enables us to enhance the current methodologies to further application.

The present chapter presents development of an advanced substructure hybrid earthquake response test procedure capable to analyze seismic performance of in-ground structures as an example of civil engineering structures. Section 2 introduces a seismic testing facility available for the advanced substructure hybrid test system at CRIEPI. In Section 3 and 4, general concept and development of our system are respectively explained. Then, Section 5 discusses comparison between a hybrid simulation and a shake table test. In addition, future scope and development of the advanced substructure hybrid testing system are briefly addressed in Section 6. Finally, the Section 7 summarizes main points in the present chapter.

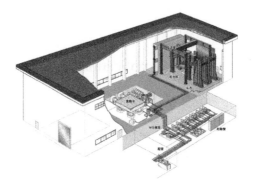

Figure 1.   Seismic testing facility at CERL, CRIEPI.

Table 1.   Specifications of the shale table.

Items	Performance description
Shaking direction	Horizontal direction
Table dimension	5 m by 5 m
Maximum mass weight	60 t
Maximum acceleration	1 G (1 m/s^2)
Maximum velocity	1.5 m/s
Maximum displacement	0.5 m
Frequency range	DC-50 Hz

2   TESTING FACILITY

The Central Research Institute of Electric Power Industry (CRIEPI) was established in 1951 as a non-profit comprehensive research organization for the utility industry in Japan. Emphasis is placed on the research for natural hazard mitigation and maintenance of social infrastructure at The Civil Engineering Research Laboratory (CERL) located in Abiko-shi, Chiba. Here, natural hazards involve wind, flood, tsunami and earthquake.

CERL has seismic testing facility completed in 2006 as depicted in Figure 1. This includes a hydraulic uni-axis (horizontal direction) shake table, hydraulic actuators, a reaction wall, a strong floor, a hybrid testing control system and data acquisition system. Major specifications of the shale table and the hydraulic actuators are tabulated in Table 1 and 2, respectively. The reaction wall, a prestressed concrete wall, has 9 m in width, 7 m in height and 2 m in thickness, while the strong floor, a reinforced concrete slab having a basement floor supported by PHC piles, has 9 m in width and 15.5 m in length and 1.5 m in thickness. As far as the data acquisition system is concerned, it mainly includes an A/D converter with 16 bit-resolution, a signal conditioner capable to DC and strain amplification for 128 components. Data acquisition operation and analysis are executed by respective PCs. In addition, a three-dimensional finite element analysis code TDAP III is prepared for the hybrid testing as will be described in detail later. TDAP III can handle various types of civil engineering structures utilizing beam, solid and spring elements associated with nonlinear hysteresis rules.

The most effective research method to demonstrate seismic-induced inelastic deformation of an in-ground reinforced concrete structure, which subjects clear soil structure interaction, may be a combined use of a large-scale shake table test and a laminar soil box as possible as they are large as shown in Figure 2. Figure 2 indicates the shake table test that was conducted at National Research Institute for Earth and Disaster Prevention in 1999 (Ohtomo et al., 2003). Although such a test can be persuasive in view of a scale and a viable understanding, testing with a limited soil box boundary

Table 2. Specifications of the actuators.

	Dynamic		Static	
Maximum load (kN)	500	100	1,000	300
Maximum velocity (m/s)	0.1	0.2	0.005	0.005
Maximum displacement (m)	0.1	0.1	0.1	0.1
Number (s)	2	1	1	3

Figure 2. Large-scale shake table test on seismic performance of an in-ground structure at NIED.

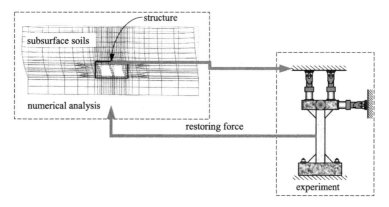

Figure 3. General concept of an advanced substructure hybrid seismic response test.

poses merely testing a model ground and structure system. This drawback yields to accept scale effect of a reinforced concrete member that has serious impact on cracking initiations or rebar yielding or plastic hinge area extension. A shake table test even in dramatically huge in size is, therefore, insufficient to study deep in ultimate seismic performance of soil structure interaction during a severe ground shaking.

3 GENERAL CONCEPT

The advanced substructure hybrid seismic response test procedure we have developed (Sakai et al., 2005) constructs a finite element model that includes subsurface soils and an underground structure as illustrated in Figure 3. In this respect, the accuracy and reliability of numerical analysis for nonlinear ground response have been well verified with a numerical correlation on centrifugal tests including soils and available strong ground motion observation records. This ensures us to assign soils to a numerical analysis part of a substructure hybrid seismic test model.

The one member whose nonlinear hysteresis behavior affects global response of a structure and cannot be mathematically modeled can be treated as an experimental part, a specimen as indicated in Figure 3. For example, a central wall can be dealt with a real size pseudo-static loading experiment of a reinforced concrete member. Then, a control computer can apply relative displacement on the top of the member using hydraulic actuators, which is resulting from a seismic response governed by soil structure interaction, subsequently a restoring force from the member is sent as a signal to the numerical analysis part.

4   DEVELOPMENT

The developed substructure hybrid experiment system employs a nonlinear finite element analysis. In doing so, major issues that must be devised are identified as follows: 1) development of stable direct time integration scheme in a numerical analysis and 2) numerical representation of a linear stiffness matrix with respect to a specimen in conjunction with a numerical analysis.

4.1   *Direct time integration scheme*

Integration algorithm with the α-OS (operator-splitting) method (Nakashima et al., 1993) was utilized. The α-OS method is referred as the mixture of the α method (Hilber et al., 1977) that successfully suppresses vibration associated with higher modes in response of multi-degrees of freedom system and the OS method that is explicit and ensures unconditional stability for larger integration time intervals.

The OS method divides the nonlinear stiffness K_{n+1} of an entire object of assessment into a linear stiffness K^0 independent of a history and nonlinear stiffness ($K^E_{n+1} = K_{n+1} - K^0$) dependent on the history. Figure 4 illustrates how to estimate the nonlinear stiffness using the restoring force and the predictor and the corrector displacements. An implicit time integral approach that is unconditionally stable, for example, the Newmark β method having the parameter α is applied to the linear stiffness. An explicit integral approach, for example, a predictor-correct method is applied to the nonlinear stiffness term as illustrated in Figure 4.

The current α-OS method expresses equilibrium attained at current time step. In an actuator-computer online test, the α-OS method is suitable to a case where only the real model portion of the object of the test has the nonlinear stiffness and the remaining numerical model portion has the linear stiffness. This scheme is unsuitable for a case where the numerical model portion also exhibits the nonlinear stiffness such as soil nonlinearity in a finite element model.

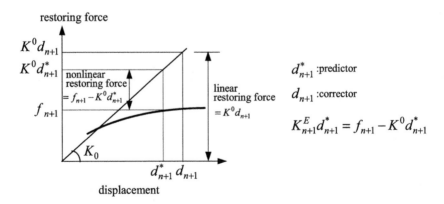

Figure 4.   Estimation of nonlinear stiffness based on the α-OS method.

If the numerical model portion has a nonlinear element, introduction of a nonlinear finite element method to solve of an equation providing a displacement made by the numerical model portion during an actuator-computer online test would prove effective. However, formulation making it possible to directly employ the combination of a nonlinear finite element method in the actuator-computer online test has not been established to date. In the present study, formulations are achieved in order to combine the α-OS method and the nonlinear finite element method. For solving this issue, the unknown parameters like displacement, velocity and acceleration are represented by an increment form like increment displacement $d_{n+1} = d_n + \Delta d n$.

4.2 *Linear stiffness matrix representation*

Let us consider a simple example of an actuator-computer online test as shown in Figure 5. Hereinafter, a specimen portion in Figure 5 is regarded as a real model portion. Referring to Figure 5, one actuator can provide a displacement to the specimen in the x direction. The linear stiffness of the specimen, a matrix K^0 as shown in Figure 4, should be precisely and readily assessed during an actuator-computer online test. This leads to improvement in testing effectively and precision.

The specimen portion as shown in Figure 5, modeled with numerical values representing the element of the beam according to the finite element method, is divided into n points. Thus, a numerical model of the specimen portion is produced. Considering that the number of portion n and the degrees of freedom m at respective point (node), the linear stiffness matrix having $n * m$ elements in rows and columns and representing the detailed model of the specimen is produced.

In an actuator-computer online test, one specimen portion is represented with two points of a prescribed portion and an applied portion and behaves as if it were formed with one element. If the number of degrees of freedom at each points, the number of axes along with the specimen portion can be vibrated at each portion m', the specimen point is treated as a virtual element having $2m'$ degrees of freedom.

In Figure 5, since the specimen portion can be deformed on axis defining the x direction, the specimen portion must be treated as a two-degrees of freedom in the actuator-computer online test. The reduction means with respect to the degrees of freedom uses the linear stiffness matrix, which represents the detailed model of the specimen. To calculate a reduced linear stiffness matrix, which represents a vibration system with the number of degrees of freedom required for the actuator-computer online test, we have applied Gyuan's static reduction method (Guyan, 1965).

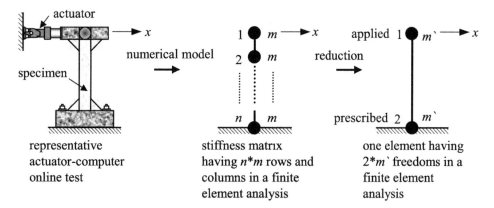

Figure 5.   General concept of linear stiffness matrix representation.

5   COMPARISON BETWEEN HYBRID SIMULATION AND A SHAKE TABLE TEST

5.1   *Method*

We have constructed a simulator, which realize the advanced substructure hybrid test system as shown in Figure 6. Control system includes four modules. They are operation module, analysis module, test module and visualization module. Modules are PCs that involve respective software applications and each module is connected via TCP/IP. The operation module controls whole test execution such as starting, interruption and resume. It also intermediates communication between analysis and test modules. The analysis module plays a global system analysis and provides computed displacement and receives reaction force from the test specimen. The test module give the computed displacement to hydraulic actuator and measure reaction force and send it to analysis module. It includes numerical simulator that is available for exercise prior to real test. Visualization module develops view plot on the PC monitor compiling analysis results and measured data.

The configuration of the shake table test for verification (Ohtomo et al., 2003) is illustrated in Figure 7. In the shake table test, a reinforced concrete box culvert (RC box culvert) having 3.0 m in width and 1.75 m in height is embedded in a laminar soil box. The soil box was filled with compacted dry sand and it was accumulated about 5 m in height. The ground acceleration history, NS component of observed ground motion recorded at Kobe University during the 1995 Hyogoken Nanbu earthquake, was applied with 50% of the observed duration.

Elastic and inelastic response associated with the soils and the RC box culvert was examined under various peak acceleration values. Ground response with 0.2% in soil shear strain under 2.25 m/s^2 in peak acceleration caused rebar yielding at the corner of the RC box culvert, while the excitation with 11.27 m/s^2 in peak acceleration resulted in 1.5% in soil shear strain and significant plastic deformation of the RC box culvert. These two cases are examined for a verification of the devised substructure hybrid test system. In addition, a full nonlinear finite element analysis is performed for evaluating the hybrid tasting performance, that is, the nonlinear finite element analysis without using a specimen and the α-OS method. Instead, only Newmark β method is employed for solving nonlinear response with direct time integration.

The numerical analysis and the specimen portions of the hybrid test system were constructed in the following manners: subsurface soils and the RC box culvert were numerically represented by a

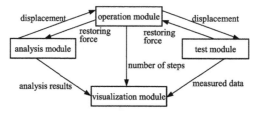

Figure 6.   Simulator for the advanced hybrid seismic response test.

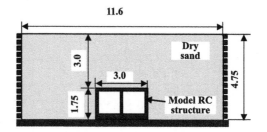

Figure 7.   Shake table test on seismic performance of an in-ground RC structure for the verification (m).

plain strain solid element and a two-dimensional beam element, respectively. A nonlinear hysteresis on stress and strain for soil elements and bending moment and curvature for beam elements were appropriately provided. A central wall of the RC box culvert was modeled as a virtual element in the hybrid test simulation as explained in the Section 4. Although a specimen in an actuator-computer online test should be loaded by an actuator, this practice was treated with a two dimension beam element having nonlinear hysteresis rule for a simulation purpose and the test module as presented in Figure 6 played this simulated experiment portion.

As far as direct time integration is concerned, time increment Δt equal to 0.01s and auxiliary parameter α equal to $-1/3$ were adopted for the hybrid simulation. On the other hand, the time integration for the full finite element analysis, Δt equal to 0.001s and auxiliary parameter β equal to 1/4 in Newmark β method is used.

5.2 *Results*

Figure 8 shows time histories of relative story displacement between the top and bottom slabs of the RC box culvert under the excitation with $2.25 \, \text{m/s}^2$ in peak acceleration, where the hybrid simulation (hybrid; the thin solid line) and the finite element analysis (FEM; the broken line) results are compared. These time histories obtained by a numerical analysis generally agree well. This verifies the validity of the α-OS method and stiffness matrix reduction scheme for the virtual specimen. Numerical analyses results well assess peak response of the relative story displacement involved in the shake table results (experiment; the bold solid line), although minor discrepancies are observed in relatively smaller amplitude response duration.

Similarly, the relative story displacement accompanied by inelastic response is discussed in Figure 9. The hybrid simulation still maintains in a sufficient accuracy in comparison with the finite element analysis results. This further intensifies the validity of the time increment form of the α-OS method as well as the Guyan's static reduction procedure. However, considerable different response between the numerical simulations and the shake table test results is observed. This is probably arising form the inaccuracy of soil nonlinear hysteresis modeling. Although the hybrid simulation poorly agree with the shake table test results under $11.27 \, \text{m/s}^2$ in peak acceleration

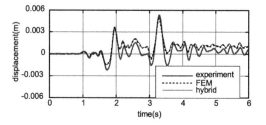

Figure 8.   Relative story displacement time histories under the $2.25 \, \text{m/s}^2$ in peak acceleration excitation.

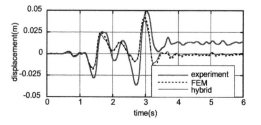

Figure 9.   Relative story displacement time histories under the $11.27 \, \text{m/s}^2$ in peak acceleration excitation.

excitation, the hybrid simulation fairly valid for soil structure interaction problem that involves significant large nonlinearity.

6   FUTURE DEVELOPMENT

Although we are promoting several earthquake resistance capability demonstration studies using the advanced substructure hybrid test system, there is still need to develop and enhance the current system. These are extension to three-dimensional hybrid testing. As is schematically presented in like Figure 3 or Figure 5, the degrees of freedom in the advanced substructure system is limited to two dimension expressed by three degrees of freedom in plane. A three dimensional-actuator-computer online test requires six degrees of freedom in accordance with an appropriate control means. In addition, a loading steel table that can harmonize six hydraulic actuators with respect to three dimensional-axes is essential. For example, the other five actuators movements must follow the applied displacement for one specified freedom. Furthermore, a more powerful computation devise using a cutting edge electronics technology should be introduced to execute an actuator-computer online test in realistic testing hours.

7   CONCLUSIONS

The present chapter presents development of an advanced substructure hybrid earthquake response test capable to examine seismic performance of in-ground structures as an example of soil structure interaction problem. Devised features, a verification of its performance using a simulator and a possible development direction can be summarized as below.

The developed substructure hybrid experiment system utilizes a nonlinear finite element analysis for the numerical analysis portion. To incorporate the nonlinear finite element analysis with an actuator-computer online test, we improved the current α-OS method so that it can be dealt with an increment form for unknown parameters such as displacement. In addition, Guyan's static reduction method was introduced to construct a linear stiffness matrix for the experiment portion. This ensures that one element can virtually represents the required number of freedoms and elements, which are usually produced in modeling of a finite element analysis, using the number of applied and fixed portions of the specimen.

Relative story displacements of the model RC culvert influenced by nonlinear soil structure interaction obtained by the large-scale shake table test were assessed by the advanced hybrid test procedure as well as a nonlinear finite element analysis. As a result, the advanced substructure test procedure and the nonlinear finite element analysis exhibit almost unified estimation for the relative story displacement. Although the incoherent response between these analyses and the shake table test results were observed particularly in duration associated with significant nonlinearity, the validity of the improved α-OS method scheme for direct time integral algorithm and Guyan's static reduction method are fairly verified.

One of the directions to enhance the current system can be identified as a three dimension version. For this purpose, testing control, a loading table and a cutting edge computer device are major areas that shoud be innovated and developed.

ACKNOWLEDGEMENT

The contents of the present chapter are a part of the cooperative study between Hitachi, Ltd and Central Research Institute of Electric Power Industry conducted from FY2001 to FY2003.The authors express their sincere appreciation for valuable suggestions and contributions made by researchers and engineers involved in this cooperative study.

REFERENCES

Guyan, R.J. 1965. Reduction of Stiffness and mass matrices, *AIAA Journal*, 3–2, 380.

Hilber, H.M., Hughes, T.J.R. & Taylor, R.L. 1977. Improved numerical dissipation for time integration algorithms in structural dynamics, *Earthquake Engineering and Structural Dynamics*, 4, 283–292.

IAEA 2007. Preliminary findings and lessons from the 16 July 2007 earthquake at Kashiwazaki-Kariwa NPP, *Mission Report Volume 1, Engineering Safety Review Services Seismic Safety Expert Mission*.

Iida, H., Hiroto, T., Yoshida, N. & Iwafuji, M. 1996. Damage to daikai subway station, *Special Issue of Soil and Foundations*, 283–300.

Mahin, S.A. et al. 1989. Psuedo dynamic test method-current status and future directions, *Journal of Structural Engineering, ASCE*, 115(8), 2113–2128.

Matsui, J., Ohtomo, K. & Kanaya, K. 2004. Development and validation of nonlinear dynamic analysis in seismic performance verification of underground structures, *Journal of Advanced Concrete Technology*, 2(1), 25–35.

Nakashima, M., Akazawa, T. & Sakaguchi, O. 1993. Integration method capable to controlling experimental error growth in substructure pseudo dynamic test (in Japanese), *Journal of Structure and Construction Engineering, AIJ*, 454, 61–71.

Ohtomo, K., Suehiro, T., Kawai, T. & Kanaya, K. 2003. Substantial cross section plastic deformation of underground reinforced concrete structures during strong earthquakes (in Japanese), *Journal of Japan Society of Civil Engineers*, 724(I-62), 157–175.

Sakai, M., Hagiwara, Y., Ohtomo, K., Matsuo, T., Dozono, Y. & Fukuyama, M. 2005. Development of visual and versatile hybrid seismic testing system incorporated with Non-linear finite element analysis, *Proceedings of the First International Conference on Advances in Experimental Structural Engineering*, 291–298.

Takanashi, K. & Nakashima, M. 1987. Japanese activities on on-line testing, *Journal of Engineering Mechanics, ASCE*, 113(7), 1104–1032.

CHAPTER 18

Hybrid simulations of nonlinear reinforced concrete frames

F. Ragueneau, A. Souid, A. Delaplace & R. Desmorat
LMT-Cachan, France

ABSTRACT: This paper aims at giving some insights within the pseudo-dynamics tests on reinforced concrete structures. Nonlinear substructuring is used making benefits of simplified finite element analysis based on multifibers beams theory approach. Continuum damage mechanics is used for the modelled structure allowing to create a realistic dynamic environment for the tested substructure.

1 INTRODUCTION

The comprehension of the ultimate behaviour of Civil Engineering structures subject to natural or industrial risks such as shocks, impacts, earthquake, explosions, ... can be handled in two ways: experimental testing and numerical modelling. In earthquake engineering, the major drawback remains in the experimental work. One has to deal with large scale structures subject to dynamic and complex loading. Classical tests are performed on shaking tables, allowing reproducing real or artificial earthquakes, but with reproducibility difficulties and physical measurement limitations. To overcome these difficulties, the pseudo-dynamic or hybrid testing are under developments (Pegon & Pinto, 2000). A combination between the numerical modelling (into which one can introduce the suitable model of material behaviour) and a test on parts of the structures can be made to better understand the structure response while benefiting from the substructuring technique. To numerically determine the inertia forces for performing static tests instead of dynamic ones leads to the so-called pseudo-dynamics (PSD) modelling. Such hybrid approaches have already been successfully adopted to treat the case of multi support bridges (Pinto *et al.*, 2004), rubber base isolators (Molina *et al.*, 2002) or to determine the damping ratio of reinforced concrete structures (Carneiro *et al.*, 2006). We describe and present the results for such PSD tests with sub-structuring technique carried out in nonlinear range, the originality being here the use of realistic anisotropic damage model for concrete. Taking into account the nonlinear behaviour of the modelled structure is of major importance since the concomitant stiffness or eigen-frequency decrease changes the whole response of the structure and so, the boundary conditions and loadings of the tested structure.

The tests were conducted for a two level reinforced concrete frame subject to a real two components earthquake (horizontal and vertical). We present the use of computations for the modelled and tested sub-structures in the PSD tests. Both an implicit and an explicit time integration schemes are used for the simulated parts in parallel with an explicit one used for the tested part. Tests results are used to identify and validate the nonlinear constitutive equations. For instance, a three dimensional damage model with induced damage anisotropy is described. The quasi-static condition of the tests allows performing refined field measurements using the digital images correlation techniques. At the scale of a reinforced concrete beam, one can distinguish for different geometries and steel reinforcement ratios, the rupture kinematics making easier numerical model identification.

2   PSEUDO-DYNAMIC TESTING

2.1   *General scheme*

Evaluation of the seismic response of a structural system is usually conducted using a shaking table. However, shaking-table experiments for large-scale structures are difficult, for instance due to table capacity limitations. An alternative way of testing full or large scale structures is the PSD testing (Shing and Mahin, 1984, Takanashi and Nakashima, 1987). The PSD testing is an experimental technique developped to evaluate the seismic performance of structure samples in a laboratory by means of computer-controlled simulation. It is an hybrid method, in which the structural displacements due to the earthquake are computed by using a stepwise integration procedure. Let us consider the dynamic equilibrium equation under seismic external acceleration $a(t)$:

$$M\ddot{u}(t) + C\dot{u}(t) + r(t) = f(t) = -M\{1\}a(t) \tag{1}$$

where M, C are the mass and damping matrices, $\dot{u}(t)$ and $\ddot{u}(t)$ are respectively the relative velocity and acceleration vectors at time t. Knowing variables at time t_n one can compute displacement and velocity at time t_{n+1} by using a numerical scheme. Only the $r(t)$ forces are experimentally measured. Numerical time discretization schemes belong to the Newmark family. Within the framework of experiment-computation interaction (Shing *et al.*, 1991), an efficient and pragmatic choice consists in implementing an Operator Splitting algorithm, allowing for a direct integration without iteration in the linear range (Nakashima *et al.*, 1993). In nonlinear regime up to rupture, it becomes necessary to damp the high frequencies, sources of numerical instabilities when an explicit procedure is adopted. In that purpose, the OS technique can be coupled to the HHT algorithm (Hilber *et al.*, 1977, Combescure & Pegon, 1997). Knowing the accelerogram and so the acceleration vector $\ddot{u}^{n+1}$, the displacements and velocities are predicted as following

$$u_{trial}^{n+1} = u^n + \Delta t \dot{u}^n + \frac{1}{2}\Delta t^2(1-2\beta)\ddot{u}^n \tag{2}$$

$$\dot{u}_{trial}^{n+1} = \dot{u}^n + \Delta t(1-\gamma)\ddot{u}^n \tag{3}$$

and corrected using:

$$u^{n+1} = u_{trial}^{n+1} + \Delta t^2\beta\ddot{u}^{n+1} \tag{4}$$

$$\dot{u}^{n+1} = \dot{u}_{trial}^{n+1} + \Delta t\gamma\ddot{u}^{n+1} \tag{5}$$

with $\beta = (1-\alpha)^2/4$ et $\gamma = (1-2\alpha)/2$. For $\alpha = 0$, the classical Newmark scheme (1/2, 1/4) is recovered and for $\alpha \in [-1/3; 0[$, the numerical scheme dissipates energy. To get an explicit solution, one may approximate the stiffness forces by:

$$r^{n+1}(u^{n+1}) \approx K^I u^{n+1} + \left(r_{trial}^{n+1}\left(u_{trial}^{n+1}\right) - K^I u_{trial}^{n+1}\right) \tag{6}$$

K^I is a stiffness matrix (from the initial virgin one to the tangential one). Using the equation of motion at time $n+1$ shifted of α, one may obtain the acceleration vector by solving the linear algebraic system:

$$\widehat{M}\ddot{u}^{n+1} = \widehat{f}^{n+1} \tag{7}$$

with $\widehat{M} = M + \gamma \Delta t (1 + \alpha)C + \beta \Delta t^2 (1 + \alpha)K^I$ and:

$$
\begin{aligned}
\widehat{f}^{n+1} = {} & (1+\alpha)f^{n+1} - \alpha f^n + \alpha r_{trial}^{n+1} - (1+\alpha)\, r_{trial}^n + \alpha C \dot{u}_{trial}^n \\
& - (1+\alpha)\, C \dot{u}_{trial}^{n+1} + \alpha(\gamma \Delta t C + \beta \Delta t^2 K^I)\ddot{u}^n
\end{aligned}
\tag{8}
$$

2.2 Sub-structuring

The PSD testing with substructuring can significantly reduce the cost of the tests to get the seismic performance of the structures (Chung *et al.*, 1999). In substructuring technique, a physical model is built only on the part or parts where nonlinearity is expected (the physical substructure), with the remaining parts modeled computationally (the numerical substructure). This method initially developed by (Takanashi and Nakashima, 1987), (Mahin and Shing, 1985) has been considerably extended by researchers at the JRC (Buchet and Pegon, 1994). The numerical part is simulated by using a finite element code in a computer connected through a network with other computers that realize the experimental procedures of the PSD test. The displacement at the interface between the physical and numerical substructures is obtained and applied to the test specimen by hydraulic actuators. The resulting resistance forces are measured by load cells and fed back to the numerical model, together with the next increment of earthquake ground motion. A new interface displacement is then calculated and applied to the tested specimen, and the loop is repeated until the test is completed, (Pegon & Pinto, 2000, Chang, 2001, Williams and Blakeborough, 2001).

We denote next by the upperscripts 1 and 2 the matrices corresponding respectively to the simulated and the tested substructure. The *b* index corresponds to the interface nodes between the two substructures. Based on the equation 7, the system to solve sums up to:

$$
\begin{bmatrix}
{}^{(11)}\widehat{M} & {}^{(1b)}\widehat{M} & 0 \\[1.5ex]
{}^{(b1)}\widehat{M} & {}^{(bb)}\widehat{M} & {}^{(b2)}\widehat{M} \\[1.5ex]
0 & {}^{(2b)}\widehat{M} & {}^{(22)}\widehat{M}
\end{bmatrix}
\begin{bmatrix}
{}^{(1)}\ddot{u}^{n+1} \\[1.5ex]
{}^{(b)}\ddot{u}^{n+1} \\[1.5ex]
{}^{(2)}\ddot{u}^{n+1}
\end{bmatrix}
=
\begin{bmatrix}
{}^{(1)}\widehat{f}^{n+1} \\[1.5ex]
{}^{(1b)}\widehat{f}^{n+1} + {}^{(2b)}\widehat{f}^{n+1} \\[1.5ex]
{}^{(2)}\widehat{f}^{n+1}
\end{bmatrix}
\tag{9}
$$

A static condensation applied to interface nodes allows to treat only two systems: the first one for the simulated substructure and the second one for the tested substructure.

In order to account for the diffused cracking in the whole concrete structure, the nonlinear behaviour of materials has to be introduced in the simulated substructure as well. In order to ensure efficiency and robustness, the framework of simplified multifibres analysis has been chosen altogether with the consideration of continuum damage.

2.3 Numerical implementation and multifibre analysis

For a simple reason of excessive computational costs, complete 3D approaches to structural dynamics in civil engineering are not commonly used. Nonlinear dynamic analyse of complex Civil Engineering structures based on a detailed finite element model require large scale computations and handle delicate solution techniques. The necessity to perform parametric studies due to the stochastic characteristic of the input accelerations imposes simplified numerical modeling which will reduce the computation cost. In classical multifibre analysis (Bazant *et al.*, 1987, Spacone *et al.*, 1996) the latter is achieved by selecting the classical Euler-Bernoulli beam model for representing the global behavior of the structural components of a complex civil engineering structure. With respect to the large spreading of the zone with nonlinear behavior it is further seek to limit the model complexity (and resulting computational costs) by limiting the diversity of possible deformation global patterns which is achieved in a multifibre beam model with fibres restricted to beam kinematics and with each one employing its own constitutive model (see figure 1).

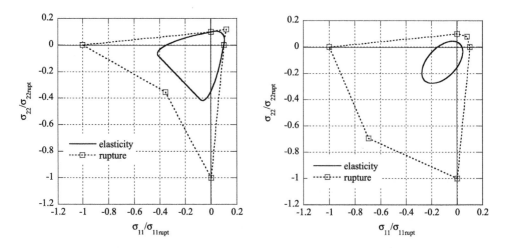

Figure 1.   (a) Elasticity and rupture for the Mazars equivalent strain. (b) Elasticity and rupture for the de Vree equivalent strain.

The main advantage of using a multifibre type finite element concerns the possibility to use a simple uniaxial behavior which allows for a very efficient implementation of nonlinear constitutive equations. Note that this is no longer possible for thick beams where shear strains play a major role (Dubé, 1994).

The multifibre beam element developed herein employs the standard Hermite polynomial shape functions to describe the variation of the displacement field along the beam. For the Euler-Bernoulli element, the shear forces are computed at the element level through the equilibrium equations. Reinforcement bars are introduced as special fibres, whose behaviour is obtained as a combination of those for concrete and steel (mixture law). The difference with "classical" beam elements concerns the cross section behaviour, i.e. the relation between the generalized strains and the generalized stresses. The chosen moduli can be initial, secant or tangent, depending upon the iterative algorithm used to solve the global equilibrium equations. The components of the constitutive matrix are computed by means of numerical integrations, often with one Gauss point per fibre. For the Euler-Bernoulli element, the shear forces are computed at the element level through the equilibrium equations (included in the Hermite polynomial shape functions).

When dealing with structures such as shear walls, which posses the slenderness ratio far from the classical beam ones, a more reliable representation of shear deformations and shear stresses has to be provided. One possibility in that respect is to use the classical Timoshenko beam model, which can describe the constant shear strain. The main difficulty of developing the finite element implementation of the Timoshenko beam model concerns the so-called shear locking phenomena, or inability of the standard finite element approximations to represent pure bending vanishing shear modes. A number of different remedies to shear locking problem has been proposed, ranging from selective or reduced integration, assumed shear strain, enhanced shear strain or hierarchical displacement interpolations. A recent work of Kotronis (2000) extends these ideas in order to construct shear locking remedies for a mulifibre Timoshenko beam.

3   ANISOTROPIC DAMAGE MODEL FOR CONCRETE

Concerning the concrete constitutive equations, a refined modelling within the earthquake engineering scope should account for decrease in material stiffness as the microcracks open, stiffness recovery as crack closure occurs, inelastic strains concomitant to damage and induced anisotropy. The latter is obtained by an anisotropic damage model based on Continuum Damage Mechanics and the present work emphasizes the fact that realistic damage models cab be used in pseudo-dynamic hybrid simulation. The model is written within the thermodynamics framework and introduces only one damage 2nd order tensor variable. To describe the damage evolution, a damage criterion of Mazars (Mazars, 1984) type is used. It introduces an equivalent strain computed from the positive part of the strain tensor. The numerical scheme used for the implementation in a F.E. code is implicit, with all the advantages of robustness and stability. However, the constitutive equations of the anisotropic damage can be solved in an exact way on an integration time step in the multifibre 1D case. The calculation of the damage and of the stress is then completely explicit from a programming point of view (Desmorat *et al.*, 2007). The consideration of anisotropic damage is more physical and more representative of the cracking pattern observed in tension and in compression. The complexity will be only mathematical (coupling with a damage tensor instead of a scalar) as one will end up with a limited number of material parameters (5 including elasticity).

3.1 *Elasticity coupled with anisotropic damage*

The damage state is represented by the 2nd order tensor $\boldsymbol{D}$ and there is one known thermodynamics potential $\rho\psi^*$ (Ladevèze, 1983) from which derives a symmetric effective stress $\tilde{\sigma}$ independent from the elasticity parameters (Lemaitre & Desmorat, 2000):

$$\rho\psi_0^* = \frac{1+\nu}{2E} Tr\left[(\mathbf{1}-\boldsymbol{D})^{-1/2}\boldsymbol{\sigma}^D(\mathbf{1}-\boldsymbol{D})^{-1/2}\boldsymbol{\sigma}^D\right] + \frac{1-2\nu}{6E}\frac{(Tr\boldsymbol{\sigma})^2}{1-Tr\boldsymbol{D}} \tag{10}$$

with E, ν the Young modulus and Poisson ratio of initially isotropic elasticity and $\boldsymbol{\sigma}^D = \boldsymbol{\sigma} - 1/3 Tr[\boldsymbol{\sigma}]\mathbf{1}$ is the deviatoric.

Quasi-brittle materials such as concrete exhibit a strong difference of behavior in tension and in compression due to damage. This micro-defects closure effect usually leads to complex models when damage anisotropy is considered (Ladevèze, 1983, Chaboche, 1993, Dragon et Halm, 1996) and the purpose next is to show that it is nevertheless important for cyclic applications. If no additional material parameters are introduced, the thermodynamics potential reads:

$$\rho\psi^* = \frac{1+\nu}{2E} Tr[(\mathbf{1}-\boldsymbol{D})^{-1/2}\boldsymbol{\sigma}_+^D(\mathbf{1}-\boldsymbol{D})^{-1/2}\boldsymbol{\sigma}_+^D + \langle\boldsymbol{\sigma}^D\rangle_-^2] + \frac{1-2\nu}{6E}\left[\frac{\langle Tr\boldsymbol{\sigma}\rangle_+^2}{1-Tr\boldsymbol{D}} + \langle -Tr\boldsymbol{\sigma}\rangle_+^2\right] \tag{11}$$

$\langle\cdot\rangle_-$ corresponds to the negative part of a tensor, expressed in its eigen-coordinates. In order to keep differentiability properties of the Gibbs free energy, the special positive part $\boldsymbol{\sigma}_+^D$ of $\boldsymbol{\sigma}^D$ has to be carefully built (Ladevèze, 1983, Lemaitre & Desmorat, 2005). The following eigenvalue problem (eigenvalues λ^I and corresponding eigenvectors $\vec{T}^I$) has to be solved:

$$\boldsymbol{\sigma}_+^D\vec{T}^I = \lambda^I(\mathbf{1}-\boldsymbol{D})^{1/2}\vec{T}^I \tag{12}$$

The normalization being defined as: $\vec{T}^{I^T}(\mathbf{1}-\boldsymbol{D})^{1/2}\vec{T}^J = \delta_{IJ}$. The positive part of the deviator is then expressed as:

$$\boldsymbol{\sigma}_+^D = \sum_{I=1}^{3}[(\mathbf{1}-\boldsymbol{D})^{1/2}\vec{T}^I][(\mathbf{1}-\boldsymbol{D})^{1/2}\vec{T}^I]^T\langle\lambda^I\rangle_+ \tag{13}$$

The elasticity law reads

$$\boldsymbol{\varepsilon} = \rho\frac{\partial\psi^*}{\partial\boldsymbol{\sigma}} = \frac{1+v}{E}Tr\left[((1-\boldsymbol{D})^{-1/2}\boldsymbol{\sigma}_+^D(1-\boldsymbol{D})^{-1/2})^D + \langle\boldsymbol{\sigma}_-^D\rangle_-^D\right]$$

$$+ \frac{1-2v}{3E}\left[\frac{\langle Tr\boldsymbol{\sigma}\rangle_+}{1-Tr\boldsymbol{D}} - \langle -Tr\boldsymbol{\sigma}\rangle_+\right]\mathbf{1} \tag{14}$$

$$= \frac{1+v}{E}\tilde{\boldsymbol{\sigma}} - \frac{v}{E}Tr\tilde{\boldsymbol{\sigma}}\mathbf{1}$$

and defines the symmetric effective stress $\tilde{\boldsymbol{\sigma}}$ independent from the elasticity parameters:

$$\tilde{\boldsymbol{\sigma}} = \left[(1-\boldsymbol{D})^{-1/2}\boldsymbol{\sigma}_+^D(1-\boldsymbol{D})^{-1/2}\right]^D + \langle\boldsymbol{\sigma}_-^D\rangle_-^D + \frac{1}{3}\left[\frac{\langle Tr\boldsymbol{\sigma}\rangle_+}{1-Tr\boldsymbol{D}} - \langle -Tr\boldsymbol{\sigma}\rangle_+\right]\mathbf{1} \tag{15}$$

The notation $\langle x\rangle_+$ stands for the positive part of a scalar, $\langle x\rangle_+ = x$ if $x > 0$, $\langle x\rangle_+ = 0$ else.

3.2   *Damage threshold function*

As for plasticity, the elasticity domain can be defined through a criterion function f such as the domain $f < 0$ corresponds to elastic loading or unloading. Many criterion can be used, written in terms of stresses such as plasticity criteria, strains, or strain energy release rate density leading or not to dilatancy in compression. The purpose here is to built a constitutive model with a restricted number of material parameters, robust and easy to implement in Finite Element computer codes. Dilatancy will not be taken into account and one will accept an open criterion for the tricompression states. These remarks lead us to the simple choice of Mazars criterion, function of the positive extensions $\langle\varepsilon_I\rangle$ of the Ith principal strain ε_I,

$$f = \hat{\varepsilon} - \kappa(tr\boldsymbol{D}) \quad \text{with } \hat{\varepsilon} = \sqrt{\langle\boldsymbol{\varepsilon}\rangle_+ : \langle\boldsymbol{\varepsilon}\rangle_+} = \sqrt{\sum\langle\varepsilon_I\rangle^2} \tag{16}$$

where $\hat{\varepsilon}$ is the equivalent strain for quasi-brittle materials and κ is the elastic strain limit in tension. Different expressions for the equivalent strain may be adopted, allowing dealing with biaxial behaviour in a more appropriate way. For example, one can consider the de Vree (de Vree *et al.*, 1995) formulation:

$$\hat{\varepsilon} = \frac{k-1}{2k(1-2v)}I_1 + \frac{1}{2k}\sqrt{\frac{(k-1)^2}{(1-2v)^2}I_1^2 - \frac{12k}{(1+v)^2}J_2} \tag{17}$$

with $I_1 = Tr\boldsymbol{\varepsilon}$ and $J_2 = 1/6I_1^2 - 1/2\boldsymbol{\varepsilon}:\boldsymbol{\varepsilon}$. The biaxial response of anisotropic modelling using Mazars or de Vree equivalent strain is given in Figure 1.

3.3   *Damage evolution laws*

To propose a damage model written in the thermodynamics framework, consider a damage pseudo-potentiel $F = Y : \langle\boldsymbol{\varepsilon}\rangle_+$ where $\boldsymbol{\varepsilon}$ acts as a parameter so that the damage evolution law is derived from the normality rule as

$$\dot{\boldsymbol{D}} = \dot{\lambda}\frac{\partial F}{\partial Y} = \dot{\lambda}\langle\boldsymbol{\varepsilon}\rangle_+ \tag{18}$$

The damage multiplier $\dot{\lambda}$ is determined from the consistency condition $f = 0, \dot{f} = 0$. with $f = \hat{\varepsilon} - \kappa(\xi)$, and, $\xi = \frac{\boldsymbol{D}:\langle\boldsymbol{\varepsilon}\rangle_+}{\max(\varepsilon_I)}$ the effective damage, effective as it only acts in the directions of the positive extensions $\langle\boldsymbol{\varepsilon}\rangle_+$.

Concerning the consolidation function κ to account for damage increase in tension as well as in compression, a possible choice is: $\kappa = \kappa(\xi)$. One has still to define the function κ. One use (Desmorat *et al.*, 2007):

$$\kappa(\xi) = a. \tan\left[\frac{\xi}{aA} + \arctan\left(\frac{\kappa_0}{a}\right)\right] \tag{19}$$

with κ_0 the damage threshold, A and a the damage parameters.

3.4 *Model responses*

Using the following material parameters ($E = 42$ GPa, $\nu = 0.2$, $\kappa_0 = 5\ 10^{-5}$, $A = 5\ 10^3$, $a = 2.93\ 10^{-4}$), the uniaxial responses in tension and in compression of the model are given in Figure 2a. The cyclic behaviour is presented in Figure 2b. The unilateral behaviour as well as the damage deactivation when passing from tension to compression are recovered. Note the low number of material parameters (5): E and ν for elasticity, κ_0 as damage threshold, a and A as damage parameters.

3.5 *Numerical implementation*

The anisotropic damage model is in fact quite simple to implement in a FE code. A global resolution of the equilibrium equations gives the displacements at time t_{n+1} with the internal damage variable $\boldsymbol{D} = \boldsymbol{D}_{n+1}$ kept unchanged from the last computed increment t_n. The strains $\boldsymbol{\varepsilon}_{n+1}$ at each Gauss point are calculated from the elements interpolation functions. To integrate the constitutive equations means to determine the stress $\boldsymbol{\sigma}_{n+1}$ and the damage $\boldsymbol{D}_{n+1}$ at time t_{n+1}. An iterative process, not described here, made of global equilibrium resolutions followed by local time integration of the constitutive equations often takes place. One focuses here on the numerical scheme for the local integration of the damage law.

Compute the equivalent strain (Mazars strain is used in the following as mainly 1D behaviour is needed):

$$\hat{\varepsilon}_{n+1} = \sqrt{\langle\boldsymbol{\varepsilon}_{n+1}\rangle_+ : \langle\boldsymbol{\varepsilon}_{n+1}\rangle_+} \tag{20}$$

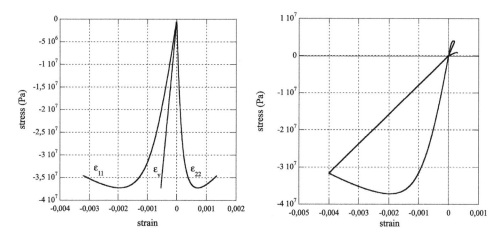

Figure 2. (a) 3D model response in compression. (b) Cyclic uniaxial model response tension—compression—tension.

Make a test on the criterion function:

$$f = \hat{\varepsilon}_{n+1} - \kappa(\boldsymbol{D}_n) \tag{21}$$

If $f \leq 0$, $\boldsymbol{D}_{n+1} = \boldsymbol{D}_n$ (material behaves elastically), else the damage must be corrected by using the damage evolution law discretized as

$$\Delta \boldsymbol{D} = \boldsymbol{D}_{n+1} - \boldsymbol{D}_n = \Delta\lambda \langle \boldsymbol{\varepsilon}_{n+1} \rangle_+ \tag{22}$$

The damage multiplier is determined from the consistency condition numerically written $f_{n+1} = \hat{\varepsilon}_{n+1} - \kappa(\boldsymbol{D}_n) = 0$ so that

$$\Delta\lambda = \frac{\boldsymbol{D}_{n+1} : \langle \boldsymbol{\varepsilon}_{n+1} \rangle_+ - \boldsymbol{D}_n : \langle \boldsymbol{\varepsilon}_n \rangle_+}{\hat{\varepsilon}_{n+1}^2} \tag{23}$$

with $\boldsymbol{D}_{n+1} : \langle \boldsymbol{\varepsilon}_{n+1} \rangle_+ = |\max\langle \varepsilon_I \rangle_+| \cdot \kappa^{-1}(\hat{\varepsilon}_{n+1})$ being known and the exact actualisation of $\boldsymbol{D}$:

$$\boldsymbol{D}_{n+1} = \boldsymbol{D}_n + \Delta\lambda \langle \boldsymbol{\varepsilon}_{n+1} \rangle_+ \tag{24}$$

Stresses computation : Using the elasticity law allow for the computation of the effective stress tensor ($\boldsymbol{E}$ is the elastic Hooke tensor),

$$\tilde{\sigma}_{n+1} = \boldsymbol{E} : \boldsymbol{\varepsilon}_{n+1} \tag{25}$$

The stress tensor is obtained through the relation between the effective stress tensor and the stress tensor in an anisotropic framework (equation 18).

The numerical scheme is fully implicit, therefore robust, but it has the main advantage of the explicit schemes: there is no need of a local iterative process as the exact solution of the discretized constitutive equations can be explicited (Desmorat *et al.*, 2004).

4   EXPERIMENTAL TESTS ON RC STRUCTURES

A full nonlinear pseudo-dynamic test with substructuring is proposed. A reinforced concrete structure, shown in Figure 3 is considered. The clamped frame is loaded with a dynamic seismic signal applied on its foundation. The anisotropic damage model is applied for concrete material, and a nonlinear plastic behaviour is chosen for steel bars. Computational time is limited in the following by taking into account a multifibre model with the Finite Element Code CAST3M. A distributed loading mass m is applied on the upper beam and on the right middle one. A concentrated loading mass M is applied in the middle of the left beam, with $M \gg m$. The frame failure is supposed to occur after the rupture of the left beam, as a low level of damage occurs in the rest of the structure. Then, just the left beam is tested and its stiffness is obtained experimentally (Laborderie, 1991) as the response of the rest of the structure is computed. An additional condition is applied on the structure: horizontal displacements of the beam ends are equal. By the way, just a single degree of freedom is controled on experimental setup. This assumption is valid in that case, where horizontal stiffness is much greater than the vertical one. The experimental set-up for testing RC beams under cyclic three points bend test is presented in Figure 4.

A first experimental result is presented in Figure 5, it shows the vertical displacement of the centre of the beam during the first seconds of the signal. The realization of quasi-static tests enables us to perform more precise measurements than in the case of tests in dynamics. At critical states of the seismic structural response, it is possible to carry out field measurements via digital images correlation (Hild, 2002). First results of this type are presented for three moments of the response: at the beginning of the earthquake, for a maximum negative moment and for a maximum positive moment. Placed at one third of the beam, the camera makes it possible to observe cracks openings in shear. Figure 6 presents the field of horizontal displacement. Openings of cracks from 50 to 700 microns thus could be observed at various instants.

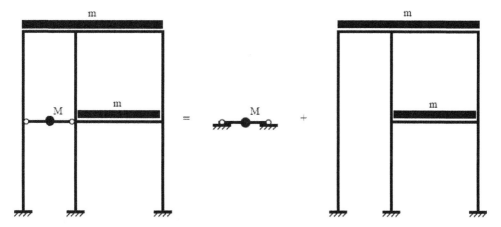

Figure 3.   Substructuring decomposition for the nonlinear testing.

Figure 4.   Experimental set-up for cyclic three points bend tests on RC beams.

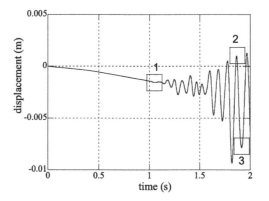

Figure 5.   Experimental mid-span deflection.

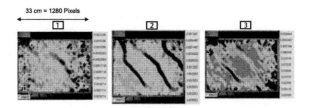

Figure 6.   Results of digital image correlation analysis performed on the tested RC beams.

5   CONCLUSIONS

Pseudo-dynamic tests with substructuring allow for dynamic studies of large structures with a moderate experimental setup. Classically inertia forces are computed and the critical part of the structure is statically tested, while the non-critical parts of the structure are modelized. Based on this approach, our work focuses particularly on the possibility to consider an elaborate damage model used for the modelized parts of the structure, and on the experimental measurements of damage on the tested structure. An anisotropic damage model, based on a second order damage tensor, allowing describing induced anisotropy and crack closure with just five parameters has been proposed. This model has been successfully used in the study of a reinforced concrete clamped frame, loaded with a seismic signal. One overloaded beam of the frame has been experimentally tested, as the rest of the structure modelized. Digital image correlation technique is used to study crack apparition and closure, allowing a fine identification and validation of the damage model. Experimental results validate the necessity to take into account the low damage level of the non-critical parts. To better introduce the physical features of the modeled structure, constitutive equations should be improved to account for the coupling between damage, hysteresis and damping. For some materials or structural elements, the time effects due to viscous behaviors have to be representative. Numerical and experimental efforts have to pushed forward to include real time testing (Haussman, 2006) altogether with nonlinear substructuring. This issue is part of collaborations between LMT-Cachan and NEES-Boulder.

REFERENCES

Bazant, Z.P., Pan, J.-Y., Pijaudier-Cabot, G., 1987. Softening in reinforced concrete beams and frames, ASCE *J. of Struct. Engrg.*, 113(12), 2333–2347.

Buchet, P. & Pegon, P., 1994. PSD testing with substructuring implementation and use, specialpublication, 1.94.25 JRC ISPRA.

Carneiro, J.O., deMelo, F.J.Q., Jalali, S., Teixeira, V. & Tomas, M., 2006. The use of pseudo-dynamic methods in the evaluation of damping characteristics in reinforced concrete beams having variable bending stiffnessn Mechanics Research Communications, 33, 601–613.

Chaboche, J.L., 1993, Development of continuum Damage Mechanics for elastic solids sustaining anisotropic and unilateral damage, *Int. J. Damage Mechanics*, 2, 311–329.

Chang, S.Y., 2001. Application of the momentum equation of motion to pseudo-dynamic testing, *Phil. Trans. R. Soc.* London, 359, 1801–1827.

Chung, J., Yun, C.B., Kim, A.S. & Seo, J.W., 1999. Shaking table and pseudo dynamic tests for the evaluation of the seismic performance of base-isolated structure, *Engng. Struct.*, 21, 365–379.

Combescure, D. & Pegon, P., 1997. α–Operator splitting time integration technique for pseudo-dynamic testing—error propagation analysis, *Soil Dynamics and Earthquake Engineering*, 16, 427–443.

de Vree, J., Brekelmans, W. & van Gils, M., 1995. Comparison of nonlocal approaches in continuum damage mechanics, *Comp. Struct.*, 55: 581–588.

Desmorat, R., Gatuingt, F., Ragueneau, F., 2007. Local and nonlocal anisotropic damage model for quasibrittle materials, *Engineering Fracture Mechanics*, 74, 1539–1560.

Dragon, A., Halm, D., 1996. Modélisation de l'endommagement par mésofissuration: comportement unilatéral et anisotropie induite, *C. R. Acad. Sci.*, t. 322, Série IIb, 275–282.

Dubé, J.F., 1994. Modélisation simplifiée et comportement visco-endommageable des structures en béton, Ph.D. thesis: E.N.S. Cachan.

Haussman, G., 2006. The CU-Boulder Fast Hybrid Test Desktop Platform, CU-NEES-06-3, University of Colorado at Boulder.

Hilber, H.M., Hughes, T.J.R. & Taylor, R.L., 1977. Improved numerical dissipation for time integration algorithms in structural dynamics, *Earthquake Engineering Structural Dynamics*, vol. 5, pp. 283–292.

Hild, F., 2002, CORRELI^{LMT}: A software for displacement field measurements by digital image correlation, Internal report N° 254. LMT-Cachan.

Kotronis, P., 2000, Cisaillement dynamique de murs en béton armé. Modèles simplifiés 2D et 3D. Ph.D. thesis: ENS-Cachan.

Laborderie, C., 1991, Phénomènes unilatéraux dans un matériau endommageable, Ph.D. thesis University Paris 6.

Ladevèze, P., 1983, On an anisotropic damage theory, Proc. CNRS Int. Coll. 351 Villars-de-Lans, Failure criteria of structured media, Edited by J.P. Boehler, pp. 355–363.

Lemaitre, J. & Desmorat, R., 2005, Engineering Damage Mechanics: Ductile, Creep, Fatigue and Brittle Failures, Springer.

Mahin, S. & Shing, P.B., 1985, Pseudodynamic method for seismic testing, *Struct. Eng.*, 111, 1482–1503.

Mazars, J., 1984, Application de la mécanique de l'endommagement au comportement non linéaire et 'a la rupture du béton de structure, Thèse d'état Université Paris 6.

Molina, F.J., Verzeletti, G. *et al.*, 2002. Pseudodynamic tests on rubber base isolators with numerical substructuring of the superstructure and strain-rate effect compensation, *Earthquake Engng. Struct. Dyn.*, 31, 1563–1582.

Nakashima, M., Akazawa, T. & Sakaguchi, O., 1993, Integration method capable of controlling experimental error growth in substructure pseudo-dynamic test. *Journal of Structural and Construction Engineering AIJ*, 454, 61–71.

Pegon, P. & Pinto, A.V., 2000, Pseudo-dynamic testing with substructuring at the elsa laboratory. *Earthquake Engng. Struct. Dyn.*, 29, 905–925.

Pinto, A.V., Pegon, P., Magonette, G. & Tsionis, G., 2004. Pseudo-dynamic testing of bridges using non-linear substructuring, *Earthquake Engng. Struct. Dyn.*, 33, 1125–1146.

Shing, P.B. & Mahin, S.A., 1984, Pseudodynamic method for seismic performance testing: theory and implementation. *UCB/EERC-84/01, Earthquake Engineering Research Centre*, University of California, Berkeley.

Shing, P.B., Vannan, M.T. & Cater, E., 1991, Implicit time integration for pseudodynamic Pegon tests. *Earthquake Engineering Structural Dynamics*, 20, 551–576.

Spacone, E., Filippou, F.C. & Taucer, F.F., 1996, Fiber Beam-Column Model for Nonlinear a Analysis of R/C Frames. I: Formulation. *Earthquake Engineering and Structural Dynamics*, 25(7), 711–725.

Takanashi, K. and Nakashima, M., 1987, Japonese activities online testing, *Engng. Mech.*, 113, 1014–1032.

Williams, M.S. & Blakeborough, A. 2001, Laboratory testing of structures under dynamic loads: An introductory review., *Phil. Trans. R. Soc. Lond.*, 1651–1669.

CHAPTER 19

Hybrid testing in aerospace and ground vehicle development

H. Van der Auweraer, A. Vecchio, B. Peeters, S. Dom & P. Mas
LMS International NV, Leuven, Belgium

ABSTRACT: The drive for innovation, cost control and time-to-market reduction has forced the ground and aerospace vehicle development process to adopt a Virtual Prototyping approach using advanced simulation methods. On its turn, this evolved the role of Physical Prototype Testing from final product refinement to supporting the CAE process. The opportunity to combine the strength of both approaches, CAE and Test, has led to a unique leverage in the form of a Hybrid Simulation methodology. This approach appears in several forms, from completing, calibrating and updating of CAE models by Test, over integrating Test-based data with CAE models into Hybrid models, to the concurrent simulation of CAE and real-world systems, allowing the virtual integration of models and systems on hardware level.

1   PROBLEM FORMULATION

1.1   *The virtual prototype engineering paradigm*

The aerospace and ground vehicles industry faces the competitively critical but conflicting demands to come up with more innovative designs and get them to the market before anyone else, developing better products in a shorter time and at a lower cost. A major step hereto was the shift towards a "Digital" design approach. Most companies have adopted an all-digital development environment for design (Computer-Aided-Design or CAD), covering the "form-and-fit" stages of the process in a "virtual space". Similarly, numerically controlled machining, robots, and a direct link of manufacturing with CAD models allow a Computer Aided Manufacturing (CAM) process. Many companies furthermore heavily invest in PDM (Product Data Management) systems and explore collaborative business models.

But next to knowing how a product looks like and how the components fit together, it is as important to get the design perform as expected for the functions required by the product's mission. Typical performances for air- and ground vehicles are propulsion efficiency, noise and vibration, reliability, safety, emissions. Many of these relate to customer requirements, and are hence competitively critical in view of the product's brand image. But they are also increasingly imposed by legislation. To take these properly into account in the design is a complicated process as they are dependent or even in conflict with each other (e.g. in relation to weight).

Traditionally, these performances were dealt with late in the development process, performing product refinement on physical prototypes, more in a way to troubleshoot problems, than as true design targets. Several advanced experimental procedures were developed hereto. But at that late stage, many development gates have been passed and the main design decisions are frozen, leading to costly, suboptimal, palliative solutions. Recent evolutions towards the use of numerical models resulted in a Virtual Prototype Engineering paradigm based on simulation tools. Detailed electrical, mechanical and other multiphysics modeling capabilities allow simulating the various performances and adapting the design to meet prior set targets. Examples are the many analytical, system-theoretic, structural finite element, vibro-acoustical, multibody, aero-acoustics, durability, thermal etc. simulations which are performed for each design.

The objective is to extend this process, leading to development cycles expressed in months instead of years. This can only happen through frontloading the engineering of the critical product qualities, using upfront concept analysis, cross-disciplinary model based product optimization and performing in-depth testing only on a reduced number of physical prototypes. Figure 1 shows the relative effort (now and desired) in each stage of the development process.

1.2   *Combining test and simulation to deliver engineering innovation*

While a purely digital design is the ambition, a fully virtual approach is not yet realistic. Insufficient calculation speed and performance of solvers is only part of the explanation since important breakthroughs in terms of computing power, parallel processing and optimized algorithms were made. Missing knowledge on exact material parameters, lack of appropriate models for complex connections, or insufficiently accurate model formulations remain major bottlenecks. The required optimization process is too complex, covering too many interrelated unknowns. Hence a combined use of test and simulation is adopted, allowing to solve engineering problems not only faster, but also more accurately compared with exclusive use of one or the other. This is illustrated in Figure 2, the Y-axis showing the required technical capability for an engineering task, the X-axis the time needed to complete the task.

The "Test Only" curve shows how the task is completed with traditional test-based methods. The "Simulation" curve shows how with simulation, part of the task can be done faster, but in

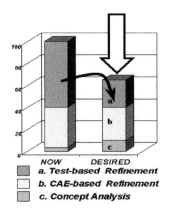

Figure 1.   Innovation targets.

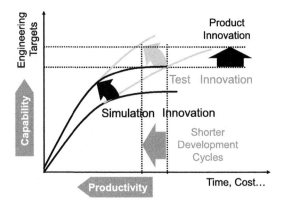

Figure 2.   Combining test and simulation to deliver innovation.

general not completed. As the required performance is available with the traditional method, test can take over where simulation reaches its limits. However, the combination of test and simulation not only delivers the required technical capability, but exceeds it. To adopt such hybrid test- and simulation approach, the total development process has to be reconsidered in view of what is feasible at which stage of the development process. At each stage, Test data and models contribute to increase accuracy and even speed up the process. The appropriate use of experimental data and experimentally obtained models on existing systems and their integration with numerical models for new designs results in a true "Hybrid" simulation approach.

2   HYBRID TESTING AND SIMULATION

Test data and models can improve and/or complement simulations in many ways. Examples range from system and component target setting through benchmark analysis, over load measurement and identification of critical model parameters, to component, sub-system and system model verification and updating (Van der Auweraer, 2004). The main added value however is to be gained from the combined use of testing and simulation results in integrated "Hybrid" system models.

A number of examples will be discussed, covering Hybrid load analysis, Hybrid structural models, Hybrid vibro-acoustic models, Hybrid acoustic radiation models, Hybrid aero-elastic models and finally Hardware-, Software- and Model-in-the-loop testing.

2.1   *Hybrid load analysis*

Simulations typically provide predictions for the response (acoustic, vibration, motion. . .) for a "unit" load, acting at a single degree of freedom. While providing qualitative insight in the intrinsic system behavior, this does not help the design to fit the quantitative targets. Understanding what the critical operating loads are, and describing these in terms of location, level and spectrum, is hence essential to perform meaningful response predictions and a true system optimization.

While simulation methods are explored to this purpose (e.g. combustion/structural models in a car engine, aero-elastic loads on an aircraft), targeted tests on existing products (e.g. predecessor vehicles) are the most reliable (and often the only possible) way to obtain this information. Dedicated transducers and measurement systems have been developed hereto such as 6-DOF wheel force or optical wheel position transducers. In many cases however, the external loads cannot be measured directly and dedicated indirect procedures (often involving partial numerical models) must be used to identify and characterize the main (critical) loads.

An example is the identification of the static and dynamic loads on flap- and slat-tracks in aircraft wing sub-systems. Strain measurements performed at accessible track parts are combined with a unit load FE model for the track (Figure 3a), to inversely identify the external loads at reference locations (e.g. leading edge). With these loads, the forward calculation to stress distribution and fatigue life within the component can be performed (Carmine, 2003).

In case no prototype is available, loads must be obtained on a predecessor or variant product. An important question is then the invariance (or known dependency) of these loads with respect to the design adaptations made. Examples are the derivation of "road profile" inputs to durability models leading to a "Virtual Test Track" and the internal engine forces for a given engine type (Bäcker, 2005). Often scaling of these quantities is used to emulate adaptations in the loading subsystem, but the extrapolation of the obtained loads (and their invariance) is limited to a given range of related vehicles. Using an inverse numerical model of the tested vehicle, these loads are derived and used as inputs for numerical simulations on the new vehicle (Figure 4).

2.2   *Hybrid structural models*

While low-frequency FE models tend to adequately describe the actual structural dynamics, this is far from true in the higher frequency range and for built-up structures with complex connections.

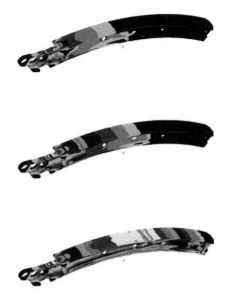

Figure 3a.   Unit load FE model, stress distribution for different slat track positions.

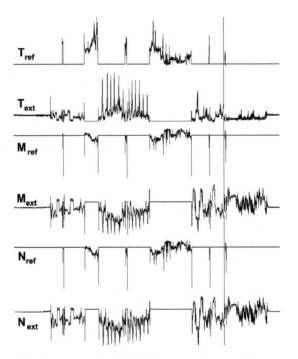

Figure 3b.   Recalculated leading edge loads (T, M, N) for two slat track positions.

Experimental data may provide critical model values such as material parameters, structural or acoustical damping, acoustic absorption, impedances etc. Specific test procedures are used to determine these. Also the interconnection between sub-structures is extremely difficult to model. In some cases, these parameters are derived through inverse procedures or from FE validation tests, confining the updating of the model to the critical parameters.

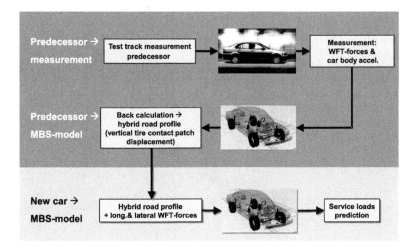

Figure 4. Virtual test track load derivation.

The real Hybrid modeling approach however combines numerical models of sub-systems or components (e.g. of a new design) with the test-models of other (e.g. existing) subsystems. With FE models, such substructuring approach is already common practice (Craig, 1968), but combining test and FE data is less straightforward. Depending on the frequency range, modal (CMS) or FRF based models are used (Duarte, 1996, Sakai, 2001), with specific requirements to the nature of the test-data needed (rigid body parameters, residual modes, local stiffness etc.).

Figure 5 shows two hybrid substructuring cases. In the first one, a vehicle engine block is modeled by FE while several added components are modeled by test data. This allows an efficient optimization of the design of the engine block making optimal use of accurate models of the other existing components. The second example concerns a hybrid aerospace model, consisting of an experimental structural model of the (existing) satellite and a detailed numerical model of an equipment rack for which the design of the interface parts is optimized.

Building such hybrid system models may involve multiple steps over a large number of subsystems. An example is the development of a hybrid road noise load and system model for the optimization of a new suspension design for an existing vehicle body (Gagliano, 2005, Jans, 2006). The inputs are the wheel spindle forces which, while not completely invariant, they are already much more vehicle independent than the body interface forces (Figure 6).

A hybrid full vehicle model can then be built (Figure 7), based on FE models of the front and rear suspension, combined with an experimental FRF model of the body. For computational efficiency purposes, the FE models of suspension components are compacted using modal reduction techniques. These hybrid road noise models are not only built to determine the body loads but are also key in the optimization of the suspension design (e.g. bushings) to minimize these body loads.

When building such complex hybrid system models, several different actions may need to be combined. Figure 8 (Cuppens, 2004) shows a refinement of the hybrid vehicle suspension model including detailed subsystem models refined based on test data. More specifically, the model uses:

(1) A test model of the pre-loaded tire from which an equivalent FE model is derived
(2) An equivalent FE model of the damper using a dedicated test rig to derive the parameters
(3) FE models of A-arm, knuckle, strut
(4) Stiffness models of the bushings (identified from updating on a quarter suspension test-rig)

The final goal of creating a hybrid model is to have an effective means to investigate possible changes to improve the interior noise due to the engine excitation. After validating the model, a number of modifications are implemented to see what their effect would be. Typical modification,

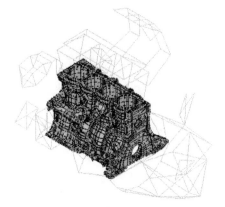

Figure 5a.   Hybrid car engine model.

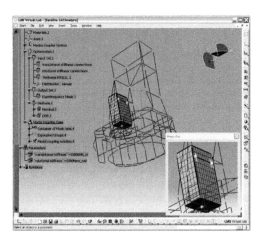

Figure 5b.   Hybrid aerospace equipment model.

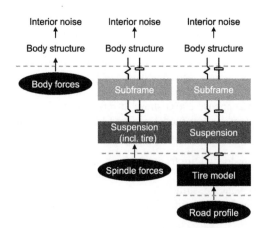

Figure 6.   Different levels of vehicle abstraction in the body interface forces identification.

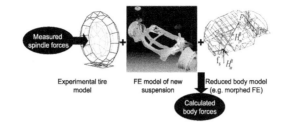

Figure 7.   Hybrid road noise model.

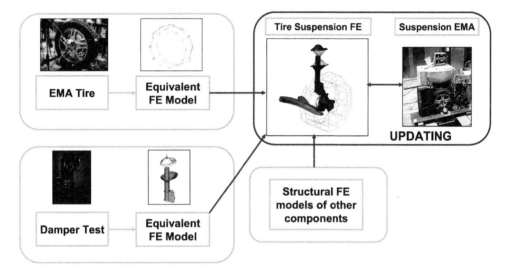

Figure 8.   Hybrid vehicle suspension model.

such as strut house thickness, subframe reinforcements etc. are combined to give typical results as are shown in Figure 9. Validation measurements on the modified system proved that the design modifications, proposed and evaluated on the model, were appropriate and that their effect could be predicted with adequate accuracy.

2.3   *Hybrid vibro-acoustic models*

The classical hybrid modeling case is this where two or more structural components are combined, but a similar approach is applicable for structural and cavity subsystems. An example is the calculation of critical panel contributions in view of interior car noise. When a trimmed-body structural FE model is available, it can be combined with experimental acoustic Frequency Response Functions (FRFs) representing the cavity, or inversely (and perhaps more realistic), when a numerical cavity model is available, it can be combined with an experimental structural model to identify the most contributing panels for a given load (e.g. engine induced high frequency vibrations) (Van der Linden, 2000). The measured FRFs describing the deformation of the structure for a given excitation are used as velocity boundary conditions of the acoustic model. Figure 10 shows an example for vehicle interior noise contribution analysis.

Using such hybrid models, arbitrary road and operational conditions can be resynthesized in real-time, providing a virtual car sound model, allowing direct evaluation of individual source or path contributions and modifications (Janssens, 2002).

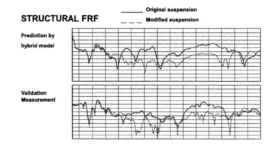

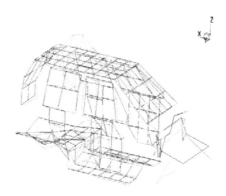

Figure 9.   Modification definition & predictions.

Figure 10a.   Hybrid vibro-acoustic model: structural test modes.

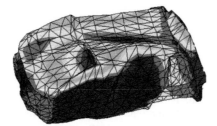

Figure 10b.   Contribution to pressure at driver's ear using a cavity numerical model.

2.4 *Hybrid acoustic radiation models*

Also in the case of an acoustic radiation problem, inverse numerical models can be used to identify acoustic sources and, together with a forward calculation, to predict acoustic radiation responses. The Boundary Element (BE) method is a numerical analysis technique, traditionally used for prediction of radiation from vibrating structures. The usual input for calculations is the vibration field at the boundaries known from structural calculations or measurements and the resulting output is the calculated sound field. The inverse problem arises when the analyst must determine the surface vibration of an existing structure, while the radiated sound field can be readily obtained from a detailed sound field mapping (e.g. when the use of accelerometers or laser vibrometers is not feasible due to the complex shape and/or accessibility of the structure). The inversion of the standard radiation calculation then offers a viable solution, generally referred to as "inverse boundary element method".

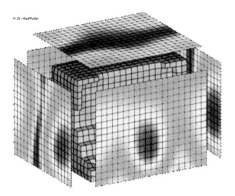

Figure 11a.   Measured pressure fields.

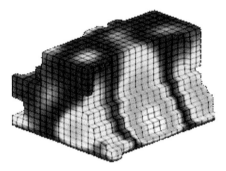

Figure 11b.   Recalculated surface loads.

In mathematical terms, instead of solving a partial differential equation for given boundary conditions, inverse BEM is a procedure in which the solution of the governing equation is known and the boundary conditions are sought. Since the BEM transforms the differential equation into an integral equation which in turn is expressed in matrix equation form, this requires a matrix inversion (Veronesi & Maynard, 1989). Specific procedures were developed for inverting the often ill-conditioned transfer matrix (regularization, use of Singular Value Decomposition (Augusztinovicz, 1999)) and to reduce the large computational effort (e.g. using Acoustic Transfer Vectors-ATV (Tournour, 2002)). Figure 11 shows the example of engine noise measured in several planes around a truck engine (Tournour, 2003). Applying an inverse numerical (ATV-based) model, the acoustic loads at the engine surface can be calculated.

From these surface loads, radiation predictions at arbitrary locations can be made.

2.5 *Software- and hardware-in-the-loop models*

The requirement to model increasingly complex systems, composed of heterogeneous components implies that model types of different natures need to be connected. This for example calls for an integration approach between discretized-geometry-based models (FE models, Multibody models), also referred to as 3-D models, and multi-physics system-theoretic models, also referred to as 1-D or 0-D models. While 3-D models are usually resulting from deterministic approaches that discretize and numerically solve complex problems such as structural dynamics and kinematics, 1-D models provide additional capabilities in simulating functional performances of complex devices (e.g. hydraulic systems, actuators, electric motors. . .) and processes (e.g. combustion, thermal, flow. . .). The combination takes place on the level of the time integration of the state equations.

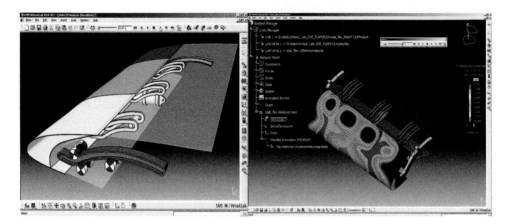

Figure 12.   3-D slat model—kinematics chain (left) and load analysis (right).

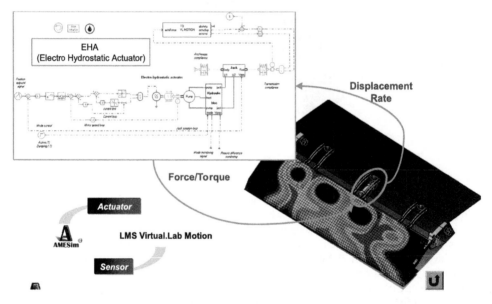

Figure 13.   Integrate full system model combining 1-D and 3-D models.

This can be through embedding the differential equations of one model into these of another model, or through a full co-simulation approach.

The task can be made even more complex embedding control laws into the multi-physics simulation, paving the way to the so called Software In the Loop applications. The simulation of the performance of aircraft and spacecraft flight control systems would not be possible without such models. Typical examples in ground vehicle design are the models for drive-by-wire vehicle systems, active suspensions, active safety systems, engine control systems. The models can be used to concurrently optimize the structural and the component design values as well as the control laws and control parameters.

This approach also opens the way to link virtual models to actual hardware systems on a physical test-bench where a model component undergoes specific testing (e.g. an Antilock Braking System). The virtual model can then represent the static and dynamic performance of the rest of the vehicle system (suspension, body. . .). This is referred to as "Hardware in the Loop testing". Particular challenges arise in relation to model calculation speed, the calculation of the virtual model having to keep pace with the hardware dynamics. The equations of the virtual model have to be converted into self-contained code which can then be compiled to run on the specific controller hardware.

Due to the complexity of the model, it may be required to reduce it in size to a lower-order model to maintain real-time performance. The development of such real-time virtual models for complex systems is definitely an area of important research for the years to come.

An example of application of the co-simulation approach is the control of the actuation of the earlier discussed aircraft slat track. From a conceptual geometry definition (e.g. CAD drawings) a multi-body model is derived that accounts for all mechanisms and represents the kinematics of the high-lift device (Figure 12). In a second step a detailed load analysis can be performed for both hypotheses of a rigid and flexible representation of the slat surface.

For modeling the functional performances, the slat track must be simulated including the physics of the actuator and its control law, requiring the execution of a full system mechatronic simulations (Figure 13). To this end an LMS Imagine.Lab 1-D model of the Electro-Hydrostatic Actuators and its control is embedded in the Virtual.Lab Motion model.

3 CONCLUSIONS

The pressure on shifting product performance optimization to earlier stages of the development process has been answered by a revolution in computer aided engineering methods, resulting in a virtual prototype engineering approach to product development. But contrary to the belief that this would eliminate testing from the development process, this has resulted in new demands for testing that are more stringent than ever. While testing was traditionally executed in a context of "Test-Analyze-and-Fix" strategy, experimental data collection and analysis is now integrated throughout the various phases of the virtual product development process, from benchmarking and target setting, through component model validation, to the establishment of true Hybrid product models. Test data are not any longer retained in isolated islands of excellence, but they play an essential role throughout the development process and are hence used throughout the extended organization. This requires that each of the applied test-procedures (operational data collection, modal analysis, acoustic testing, noise source identification, etc.) has to be critically revisited in view of the new requirements of virtual prototype refinement.

ACKNOWLEDGEMENTS

The ideas presented in this paper are the result of many discussions with colleagues and customers. Their input, including the various real-world application examples, is very much appreciated. The research on novel techniques and system solutions is conducted in the context of the EUREKA project FLITE, IWT project AIRSIM and Analysis Leads Design and EC Project InMAR. The financial support of IWT Vlaanderen and the European Commission, DG Research, is gratefully acknowledged.

REFERENCES

Augusztinovicz F., Marki F., Granat J., Hendrix W., Van der Auweraer H. 1999. Development of an Inverse Boundary Element Technique for Partial Noise Source Identification of Tyres. *Proc. Internoise 99*, Fort Lauderdale (FL-USA): pp. 1413–1416.

Bäcker M., Langthaler Th., Olbrich M., Oppermann H. *"The hybrid road approach for durability loads prediction"*, SAE paper 2005-01-0628.

Carmine R., de Voghel R., Van der Linden G., Guillaume P. 2002. Numerical life prediction of a slat track under test flight conditions. *Proc. MSC Aerospace Conference*, Toulouse (F).

Cuppens K., Mas P., Van Herbruggen J., Tatsuya O. 2004. A Hybrid Full Vehicle Model For Structure Borne Engine Noise Prediction. *Proc. SIA 2004*.

Craig R.R. Jr. 1968. Coupling of substructures for dynamic analysis. *AIAA Journal*, Vol. 6. No. 7: pp.1313–1319.

Duarte M., Ewins D. 1996. Improved experimental component mode synthesis with residual compensation based on purely experimental results. *Proc. IMAC 14*, Detroit (MI, USA): pp. 641–647.

Gagliano Ch., Martin A., Cox J., Clavin K., Gérard F., Michiels K. 2005, "*A hybrid full vehicle model for structure borne road noise prediction*," Proceedings of the 2005 SAE Noise & Vibration Conference, paper 2005-01-2467.

Gerard F., Mas P. 2005, Full vehicle hybrid simulation applied to vehicle interior noise optimization, paper No. 2005-26-337, Proc. SAE India Mobility Engineering Congress, Chennai (In), Oct. 23–25, 2005.

Jans J., Wyckaert K., Brughmans M., Kienert M., Van der Auweraer H., Donders S., Hadjit R. 2006. Reducing Body Development Time by Integrating NVH and Durability from the Start. *Proc. SAE 2006 World Congress & Exhibition*, Detroit (MI, USA): paper 2006-01-1228.

Janssens K., Adams M., Van de Ponseele P. 2002. The Integration of Experimental Models in a Real-Time Virtual Car Sound Engineering Environment. *Proc. ISMA 2002*, Leuven (B).

Sakai T., Terada M., Ono S., Kamimura N., Gielen L., Mas P. 2001. Development procedure for interior noise performance by virtual vehicle refinement combining experimental and numerical models. *Proc. SAE N&V Conference*, Traverse City (USA): SAE paper 2001-01-1538.

Tournour M., Cremers L., Guisset P., Augusztinovicz F., Marki F. 2000. Inverse numerical acoustics based on acoustic transfer vectors. *Proc. ICSV-7*, Garmisch-Partenkirchen (D).

Tournour M., Brux Ph., Mas P., Wang X., McCulloch C., Vignassa Ph. 2003. Inverse Numerical Acoustics of a Truck Engine, *Proc. SAE N&V Conference*, Traverse City (USA): SAE Paper 2003-01-1692

Van der Auweraer H., 1994. The New Paradigm of Testing in Today's Product Development Process. *Proc. ISMA 2004*, Leuven (B).

Van der Linden P.J.G.L., Gerard F., Michiels K., Van der Auweraer H., Storer D. 2000. Body in White Panel Noise Assessment through Spatial and Modal Contribution Analysis. *Proc. ISMA-25*, Leuven (B): pp. 1361–1368.

Vecchio A., Rimondi M., Janssens K., Hovmand P., Anders F., Hybrid modelling approach to predict engine noise reduction in passenger trains, *Proc. Euronoise 2006*, Tampere (Fi).

Veronesi, W.A., Maynard, J.D. 1989. Digital holographic reconstruction of sources with arbitrary shaped surfaces. *Journ. Acoust. Soc. Amer.*, Vol. 85. No.2: pp. 588–598.

CHAPTER 20

Hybrid testing & simulation—the next step in verification of mechanical requirements in the aerospace industry

L. Ayari
Ball Aerospace & Technologies Corp., Boulder

ABSTRACT: After 50 years of successful space exploration, early decisions made on hardware verification need to be revisited. Structural verification based on experimental testing has been characterized with conservatism, and while it has worked successfully, it has been responsible for many unnecessary losses of precious hardware, wasting valuable resources and causing schedule delays. The current drive for hybrid testing presents a historical opportunity for the industry to revise the way structural verification is conducted. It has to be mentioned however, that the industry has long recognized such need and as a response, developments in the context of force limited vibration was introduced and accepted as a standard. This paper proposes the expansion of the NASA-JPL concept of force limited vibration testing to combine modeling with experimental testing in a comprehensive verification plan for future space-bound hardware. This paper makes the case for the need to make the shift to hybrid testing.

1   INTRODUCTION

The journey back to the moon has begun and the engineering implementation of the new vision for exploration is in full motion (Dale, 2006). To support of the new vision for space exploration, novel verification methods are needed to replace the current cumbersome verification process. In the next twenty years and beyond, there will be a need to regain ground when programs get hit with schedule delays and or budget overruns. Not only, hybrid testing has the potential to cut cost and speed schedule, it will also help produce more reliable hardware. However, it is now that we need to act and do the necessary developments. It is relatively a small deposit of effort that we should make right now so that we have the tools ready during production.

The verification of mechanical requirements in aerospace structures is usually done through a sequence of steps: analysis, process certification, inspection and test. With inspection and test, a great deal of time and money are spent before finding out if a design is any good (Sarafin, 1995). These steps are usually independent from one another; they work however in cycles where iterations are needed until the product is fully developed and integrated in an upper system. Tremendous merit can be found in lifting the independence through combining the analysis and test components in what is proposed as Hybrid Testing.

Hybrid testing of engineering structures started to receive attention in academia and new tools are proposed to deal with verification in many academic labs. NASA and the aerospace industry should certainly lead the way in taking the hybrid testing concept from university labs and make it a reality, i.e. a production tool. There are several obvious reasons for adopting this new technology. The potential can be immediately seen in

- hardware cost reduction,
- time saving through acceleration of hardware design, and
- Providing realistic assessment of structural characteristics of large systems where neither testing nor analysis independently provide the necessary information.

The good news is that we do not have to start a new effort. It is a matter of bringing together the JPL led effort and methodology in introducing "Force Limited Vibration Testing" with the academic work of the NEES center of the University of Colorado on structural partitioning.

The aerospace industry has a strategic interest in developing new "Hybrid Testing" hardware that is capable of testing and analyzing, and therefore verifying, a test-end item in a setup where the surrounding environment is analytically and/or physically simulated concurrently (see e.g. Haussmann, 2006). In particular, such hardware needs to be able to

1. Perform classical structural testing (i.e. what shakers usually do: harmonic tests, random vibration tests, sine sweep tests, modal surveys and others.)
2. Deal with a physical test end-unit and simultaneously account for an absent subsystem through actuation. The participation finite element real time analytical models. During test, the physical unit (test subsystem) interacts instantaneously with a host subsystem represented analytically via a computer model (see Figure 1). The interface conditions between the physical and computational subsystems are imposed during test.
3. Perform real-time "Modal Identification."

The hardware vision for such hybrid shaker relies on integrating several technologies which are currently available:

- Six-DOF testing capability. Pure or combined (i.e. multi-degrees-of freedom) oscillatory accelerations or forces in the six degrees of freedom may be specified.
- Built-in dual acceleration/force control capability at the shaker head.
- Combined Pneumatic/Electromagnetic actuation to cover a frequency range between 1 Hz and 2000 Hz.
- Built-in analysis processor based on the discrete finite element modal synthesis.

One key component in the development of aerospace structures is the qualification and validation of hardware through analysis and testing. As it will become clear, several successful attempts were already made to incorporate the above capabilities in application tests. Achieving full integration in the hybrid sense as a production tool is a matter of will, and in the opinion of the author is eminent.

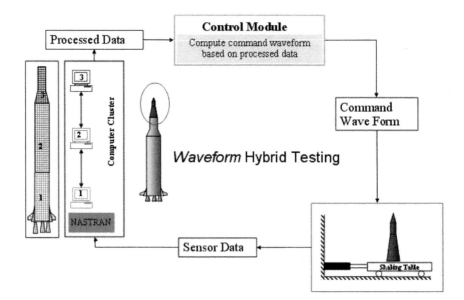

Figure 1.   Logic of hybrid testing structural partitioning.

A testing system similar to the one described above will open new dimensions in specification writing and requirement development.

2 APPLICATION OF HYBRID TESTING

Despite the fact that NASA has led the way in the research of testing methods, little attention is currently given to the subject. This waning of effort is happening despite the fact that new control technologies and new measurement tools have emerged. Interestingly, reference to encouragement to incorporate hybrid testing in research can be found in NASA's Small Business Innovation Research & Technology Transfer Program Solicitations. It is stated in the 2005 announcement for competition, under "*Advanced materials and Structural Concepts,*" (which goals are to develop high-performance materials, fabrics, modular vehicle structural concepts, and mechanical components for exploration systems):

" . . .*Modeling and structural testing techniques and analyses that support the design of modular structural concepts or their assembly are of interest. Two areas are of particular interest: one is controls-structures interaction (CSI) techniques and the second one is hybrid-test and physics based-modeling approaches. Application of advanced controls-structures interaction (CSI) techniques for measuring and controlling structural dynamics and geometry are important. Solutions for incorporation of CSI techniques for controlling such inflatable structures are also highly desirable. On hybrid modeling, ways to integrate test and physics-based models for cases where the physics-based models are not sufficient is also desirable.*"

To the best of the author's knowledge, today, there has been no serious developmental effort to produce hybrid testing equipments to meet the above recommendations. At the experimental level, the only serious development comes from the University of Colorado and the NEES program where a demonstration of a uniaxial system (for small loads) has been successful.

Overseas, England seems to be leading the way in hybrid testing. Early in 2005, the Queen inaugurated the Bristol Laboratory for Advanced Dynamics Engineering (BLADE). Blade is probably one of the most advanced dynamics research laboratories in Europe. This $35 Million laboratory includes a two-story test hall for the integrated testing of small to medium-size aero-space and mechanical engineering structures. Tests on full-size components are combined with computer simulations in a hybrid testing approach referred to as dynamic substructuring.

In South Korea, the Flight Vehicle Research Center (FRC) was established in 2004 with the support from the Korean Department of Defense. It comprises of 19 research projects conducted by 300 domestic researchers. The Lab aims to develop next-generation aerospace vehicle technologies. Research on hybrid testing is being done under the so-called virtual structural testing research program. High-speed computing capability is used through their domestic information technology infrastructure.

Work in Hybrid testing also comes from the German Aerospace Center (DLR) in collaboration with Bauhus University of Kassel where substructure testing in which the test item is considered as part of a larger system. A four DOF system was tested using an algorithm based on force compensation similar to a PID control.

In France, Hybrid testing was given a strategic priority. At the CNRS, the Laboratory of Mechanics and Technology joined force with the Center of Mathematics and their Applications (CMLA) and with the Laboratory of Electrical Engineering, Signals and Robotics in a project to develop a platform for hybrid structural testing. This common program aims at revisiting the "Validation tests" to improve model calibrations. The researchers thus hope to be able to control and improve the models of materials and structures used for a given application as well from the integrity point of view or safety as optimization of the coupled material/structure behavior. This platform which has been operational since 2002 provided new possibilities in modifying the dialogue "real/virtual." The main objective of the platform is to decrease the number of tests all while optimizing them and by capitalizing them to improve the virtual model against reality.

The platform is thought of as a modular tool, making it possible to test complex structures with arbitrary stiffness under various loading conditions. It consists of an isolated solid mass making it possible to test structures such as cars and satellites in static and dynamic conditions. Open, it will make it possible to integrate means of simulation of the environment and varied measurement capabilities. The real time processing is facilitated by the connection linking the platform and the parallel computer established in LMT-Cachan.

3   THE PROBLEM OF OVERSTRESS

In many programs where space hardware is produced, testing requirements impose a design load that is quite conservative not only making the hardware heavier, but also consuming part of the life of critical components prone to fatigue. Parts with higher frequency content are subjected to a large number of cycles from the random load profiles often imposed on the hardware. Another problem exists in the requirements that a program designs to at program initiation. In the absence of derived requirements, it is usually easy for the prime contractor to direct the subcontractor to use the General Environmental Verification Specification (GEVS) document (Milne, 1996) or apply the so-called "Mass-Acceleration Curve." By the time an estimate of the applicable loads has been generated, a heavy design has matured to a point of no return. Even the load estimates are usually very conservative. One has to admit that designers and analysts suffer during the development process for they have to analytically show the design "good" and in the same time have it fit in a small space. Engineers tend not to complain because, after all, the Customer is King. The author has experience of loads larger than later measured by one order of magnitude. In the interest of fairness, the system works; and a successful half century of flight hardware has completely changed the world. The challenge now is how to maintain a conservative approach, not waste resources and not subject the hardware to overstress.

3.1   *Design loads & environments—industry's definitions*

There has been a unification of the loads nomenclature used in the aerospace industry. Prime contractors usually furnish a set of loads to subcontractors for design purposes. At the beginning of program, and to set the work in motion, the so-called Mass Acceleration Curve (MAC) is recommended for use. This is the earliest and most conservative launch load supplied by a customer. The physical mass is the actual mass for a single degree of freedom system. For multi-degree of freedom systems, the physical mass can be approximated as the portion of mass supported by the element being analyzed. The dynamic acceleration is applied in the single direction producing the greatest load component.

The "Limit loads" are intended to provide an upper bound on the loads in primary and secondary structures. These member loads result from multiple sources such as expected launch, orbit insertion, thermal environments and handling. They are suitable for assessing the structural integrity of primary and secondary structural elements in the low to mid frequency (less than about 75 Hz). For components whose fundamental resonances are greater than 75 Hz, these limit loads may not account for vibration/acoustic/shock environments. The vibro/acoustic design requirements specify other vibration loads which are intended to assess the functional integrity of structural equipments above the 75 Hz level. However, during these tests primary and secondary structural elements are also exercised. Since these tests could cause excessive vibration response in some of the primary and secondary structure, they are "notched" (see below), if necessary, to prevent exceeding their limit loads. All primary and secondary structure, therefore, are designed for the 0–75 Hz regime limit loads and assessed to determine survivability in the vibro/acoustic/shock environments. All black-box and non-structure components are designed to survive the entire frequency regime of the vibro/acoustic/shock environments.

Launch loads are usually based on an analysis of the coupled launch vehicle/spacecraft and account for the low to mid-frequency (approx. less than 75 Hz) launch vehicle induced dynamic loading in all primary and secondary structures. The maximum loads of the launch vehicle being considered must be used. All critical launch conditions for all primary and secondary structures are checked. (e.g. ignition, liftoff, staging, shutdown, boost phases, spin up, and ignition.)

Later, and after a certain degree of maturity of the different subsystems, an "Intermediate Coupled Load Cycle" followed by a "Final Load Cycle" are generated and delivered to the subcontractor. These loads usually come lower than initial estimates and come as a relief for the parts that did not get to be shown "good" under the previous loads.

There has been a persistent problem with overstressing hardware components over the past 40 years. This problem was addressed in many contributions. Three basic methods have been devised over time to deal with overstress:

- Input Notching
- Response Limitation
- Force-limited Input

The only successful effort to overcome this problem systematically was led by JPL through the work by Scharton (e.g. Schaton, 1997) who introduced the "Force Limited Vibration Testing" technology which was received with delight from engineers from the NASA subcontracting community. This semi-empirical method is in the author's opinion the precursor to a future hybrid qualification testing methodology.

3.2 *Notching the response*

Following a modal survey, the frequencies causing unacceptable stress during tests are usually dealt with through what is called "notching." The way notching works is by creating wells in the input spectrum around the input frequencies creating stress beyond that of limit load. This method is somewhat arbitrary, in the sense it has to be done during test with not much time to study the implications of how deep the well has to be so that we do not under load or overload the test unit. Chances for causing failure or under testing are function of how well the hardware is understood through analysis.

3.3 *Response-limitation*

Through the use of critically positioned accelerometers, the test item has its response limited during test through a closed-loop control system. This process has helped tremendously in limiting the base load excitation and therefore prevented failure. The implication here is that the method is acting on the response, and not the input load, and making mistakes on where to choose the accelerometer location may cause an over-test or an under-test. Decisions are usually made on the go during test with no room for mistakes. In addition, a major problem with this method is in the fact that often the proper location where a positioning of an accelerometer is needed is inaccessible. Finally, this method has a fundamental flaw; an inherent interference between test and analysis that is keeping the door open to major errors in the verification logic.

3.4 *Force-limited vibration testing*

This is a semi-empirical method proposed by JPL in which the test-end item is mounted to the shaker through load cells which monitor the transmitted forces, and subsequently limit the load input in real time. This method extensively developed by Schaton, 1997 (Schaton, 1997) has become an industry standard, as devised in (NASA-HDBK-7004, 2004), approved by all NASA centers for use. The advent of reliable triaxial force gages has made possible a vibration-testing method based on limiting the reaction force between the shaker and the test item. This force-based limitation is introduced in a closed loop with the input shaker acceleration. The rational behind the methodology and all the technical details can be found in the NASA monograph NASA-RP-1403.

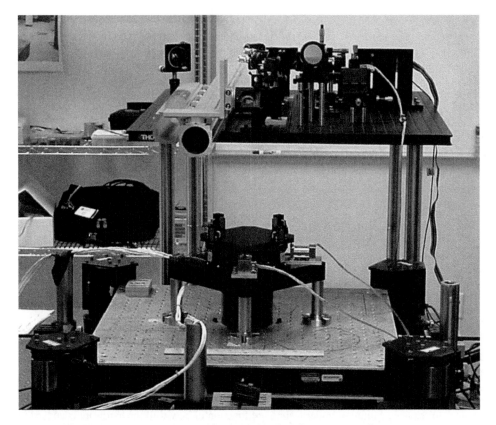

Figure 2.   Vibration input bench built at BATC, simulating jitter from a spacecraft.

4   OTHER VIBRATION INPUT & LOAD MEASUREMENT DEVICES

Advanced control strategies are becoming routine where load, acceleration and velocity sensors are designed to work in concert with clean and repeatable actuation systems. The author has some experience with such test facilities. Two vibration input benches were built at Ball Aerospace & Technologies Corporation (BATC). One was built in 1986 and one in 1996 (see Figure 2). The former is a three-axis shaker table and the latter is used as platform stabilization and was named VIB short for Vibration Input Bench.

The VIB system provides pitch, roll and vertical base motion to a four square foot optical bench that can accommodate payloads up to 50 lbs. The table stroke specifications are $+/-0.25$ in vertical and $+/-20$ mrad rotational, with vertical accelerations from 0.33 g's (fully loaded) to 0.72 g's (unloaded). The VIB system is used to simulate aircraft or spacecraft vibration spectra, testing base motion disturbance rejection and characterizing dynamic measurement instruments.

BATC also built a six-axis force measuring system (see Figure 3), called Dynamometer. The setup for the dynamometer consists of 3 orthogonal 3-axis load cells (Kistler type 9067), 9 dual mode charge amplifiers (Kistler type 5010B0), and a PC running LabView.

The associated LabView virtual instrument resolve the local load cell readings into the 3 forces and 3 moments acting on the table and then performs a Fourier transform to obtain the frequency response of the moments and forces. The load cells are oriented in the same direction and form an equilateral triangle. The outputs of the charge amps are tied to a National Instruments type BNC-2110 connector block and a National Instruments type BNC-2115 connector block, which

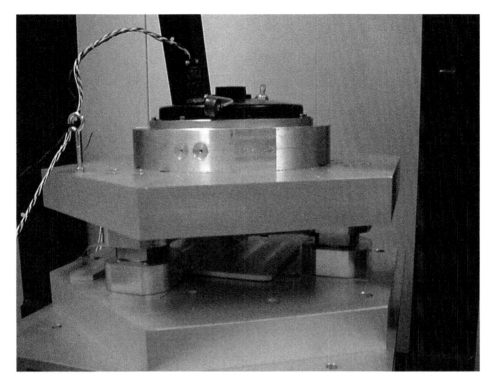

Figure 3.   BATC built dynamometer, measuring jitter isolation of the Robin momentum wheel.

connect to a National Instruments type PCI-6033E data acquisition card via a National Instruments type SH100-68-68-EP split shielded cable.

Massive load cells designed to be part of a shaker head would provide a continuous reading of the reaction loads in the six degrees of freedom at any time. Such capability can provide a mean to simulate a spring rate with certain modal characteristics in real time. It has to be noted here that there is no need for an exact reproduction of the exact interface load; it is rather the production of a boundary condition that more closely match an infinitely stiff mount that is of interest. Such boundary condition can be established through an equivalent work principle similar to what is done in classical structural analysis when complicated loads are replaced with equivalent ones based on energy balance.

5   MODEL-TEST CORRELATION AND REAL TIME MODAL IDENTIFICATION (RTM)

One of the objectives of testing hardware is to identify the modal characteristics in order to correlate with the Finite Element Model (FEM) that is used to predict loads from the combined hardware & host structure—e.g. Spacecraft (S/C) and Launch vehicle L/V coupled Loads Analysis.

The sequence of structural tests includes a formal modal survey test in which the either the S/C is hard-mounted at its base (e.g. Figure 4) and excited by electro-magnetic actuators or "stingers or a random vibration tests and the pre- and post- low-level "signature" random vibration runs are performed on a shaker table. This effectively enables the comparison of the dynamic characteristics between flight hardware and FEM. Post-test refinements to the FEM are usually conducted in order to correlate the primary modal frequencies observed during vibration tests. The primary frequencies of FEM are correlated to within 10% of the observed test primary frequencies. Primary modes are

Figure 4.   The relay mirror experiment during qualification testing.

assumed to be the fundamental modes in each of three orthogonal axes of the overall motion of a S/C that has significant effective mass. As a goal, but not as a requirement, the fundamental modes that are associated with the major appendages such as Solar Arrays, and Propulsion Modules are also correlated to within 10%. The resulting test-verified FEM is used for the subsequent Verification Coupled Loads Analysis.

The recently suggested process of real time modal testing suggested in (Lefevre et al., 2002) uses measurements of interface forces from simple low-level uniaxial sine sweep tests to first identify the so-called dynamic mass and subsequently perform modal identification taking into account parasitic motions. It is claimed that RTM allows a better piloting of the qualification runs and a simplified real time modeling updates. It has been reported in (Bricout, 2003) that several planned tests were cancelled in light of the Real Time Modal Identification procedure saving time and money.

6   SIX-DOF TESTING CAPABILITY

As mentioned earlier, the next step in testing equipment should be a hybrid shaker that integrates several well established technologies. The six-DOF testing capability delivers either pure or multi-degree-of freedom oscillatory accelerations or forces in the six degrees of freedom may be specified. It should also have a built-in dual acceleration/force control capability at the shaker head and a combined pneumatic/electromagnetic actuation to cover a frequency range between 1 Hz and 2000 Hz. Control should be delivered from a built-in processor capable of simulating input modulation based on discrete finite representation of a virtual structural system.

Figure 5 illustrates such system. The dynamics associated with such shaker has been developed at Ball Aerospace & Technologies Corporation. Central to the design is the placement of the CG of the moving mass on the pure center of rotation (CR) of the flexural system represented by flexible shells symmetrically positioned 90 degree or 120 degree symmetry). Actuation is carried

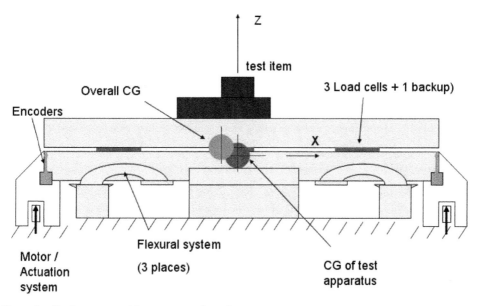

Figure 5.   Testing system with pure center of rotation.

either at three or four point. The system has position, acceleration, and force sensors that feed the information at any time.

Such initially decoupled system in the six-DOF is expected to introduce coupling due to the offset between the CG and the CR, r, once the test item is bolted on the shaker head. The nonlinear problem is solved analytically in a closed form by expanding the mode shapes and frequencies in a Taylor series. The response is then derived for any excitation as a damped and as an undamped motion. The only assumption made here is in the overall cross product of inertia of the entire system being small and its effect is negligible. Such effect would come from a non-zero cross product in the test item; the moving system of the shaker itself is balanced by design. The resulting linearized system comes with linear coupling terms in the modes shapes, while the system frequencies are found to be second order functions of the coupling components rx, ry and rz.

An example of a response due to a Z axis harmonic excitation is documented in the Appendix.

7   CONCLUSIONS

Following a critical overview of the complex task of structural qualification in the aerospace industry, it is suggested to lump some of the actual processes into a single larger one in which the key problems of overstress and other wastes get to be eliminated.

By now, it is clear that there is a fundamental need for a concept based on the emerging hybrid testing technology. Such concept which should be developed on current technology platforms. There are major technological and economical gains to be made. Some of these are:

- Solve the overstress problem,
- Eliminate many testing phases such the coupled-loads analysis work cycle,
- Provide a more realistic interface simulation,
- Eliminate unnecessary fatigue life consumption, and
- Acceleration of schedule by skipping testing of subsystems and intermediate assemblies.

In light of the strong arguments in favor of hybrid testing, making the case for this long-waited approach to qualification should not be a major challenge.

REFERENCES

Thomas P. Sarafin, (ed.) 1995. *Spacecraft Structures and Mechanism from Concept to launch*, Kluwer Academic Publishers.

Dale, S. 2006. Exploration Strategy and Architecture, *2nd Space Exploration Conference—Implementing the Vision, Houston, Texas, December 4–6, 2006.*

Haussman, G. 2006. Parallel Processing of Waveform Generation for Shaking Tables, NEES center Publication, University of Colorado, Boulder, Colorado.

Scharton, T.D. 1997. Force Limited Vibration Testing Monograph, NASA Reference Publication RP-1403.

Milne, J.S. 1996. General Environmental Verification Specification for STS & ELV, Payloads, Subsystems and Components, GEVS-SE, June 1996.

NASA-HDBK-7004 Force Limited Vibration Testing—Applicable, May 16, 2000.

NASA-RP-1403 Force Limited Vibration Testing Monograph, May 1997.

Lefevre, Y.M., Bonetti, J.C., Girard, A., Roy, N., Calvi, A. 2002. Real Time Identification Techniques for Improved Satellite Vibration Testing, *European Conference on Space Structures, Materials & Mechanical Testing, Toulouse, France, Dec. 2002.*

Bricout, J.N. 2003. Reduction of SPOT 5 Mechanical Qualification Test Program based on CLA Information and on Real Time Modal Identification During the Sine Vibrations, *S/C and L/V dynamic Environments workshop, The aerospace Corp.—El Secundo, California, June 17th–19th, 2003.*

APPENDIX

Under the application of a single harmonic loading of amplitude Po and frequency ω, the response at the combined shaker head and test-item CG is given by the vector UZf and the reactions are given by the vector RZf:

where

$$UZf_1 := 0 \qquad\qquad RZf_1 := 0$$

$$UZf_2 := 0 \qquad\qquad RZf_2 := 0$$

$$UZf_3 := \frac{P_0\phi(Wz,\omega)}{mWz} \qquad RZf_3 := -P_0\phi(Wz,\omega)$$

$$RZf_4 := P_0(-\phi(Wz,\omega) + Wz\eta(Wz,Wxx,\omega))ry$$

$$UZf_4 := -\frac{WzP_0\eta(Wz,Wxx,\omega)ry}{JxWxx} \qquad RZf_5 := -P_0(-\phi(Wz,\omega) + Wz\eta(Wz,Wxx,\omega))rx$$

$$RZf_6 := 0$$

$$UZf_5 := -\frac{WzP_0\eta(Wz,Wyy,\omega)rx}{JyWyy}$$

$$UZf_6 := 0$$

where

$$\phi(Wz,\omega) = \frac{Wz\sin(\omega\cdot t) - \omega\sin(Wz\cdot t)}{Wz^2 - \omega^2}$$

$$\eta(Wz,Wxx,\omega) = \frac{-\phi(Wz,\omega)Wxx + Wz\phi(Wxx,\omega)}{Wz^2 - Wxx^2}$$

$$\mu(Wz,Wxx,\omega) = \frac{-Wz\phi(Wz,\omega) + \phi(Wxx,\omega)Wxx}{Wz^2 - Wxx^2}$$

$$\mu(Wz,Wxx,\omega) + Wz\eta(Wz,Wxx,\omega) = \phi(Wxx,\omega)$$

$Wx, Wy, Wz, Wxx, Wyy, Wzz$ are the first fundamental frequencies of the equivalent balanced system, Jx, Jy are the inertia in the x and y axes.

Subject index